Belt Drying of Sludge

Albert Heindl

Belt Drying of Sludge

Springer Vieweg

Albert Heindl
Deining, Germany

ISBN 978-3-662-66447-6 ISBN 978-3-662-66512-1 (eBook)
https://doi.org/10.1007/978-3-662-66512-1

Springer Vieweg

This Springer Vieweg imprint is published by the registered company Springer-Verlag GmbH, DE, part of Springer Nature.
The registered company address is: Heidelberger Platz 3, 14197 Berlin, Germany

Paper in this product is recyclable.

"Sludge creates most of the problems. As raw sludge, it stands for only 1% of the treated wastewater, but it causes 30% of the costs and 90% of the headaches."
Karl Imhoff, famous German expert in wastewater treatment (1876–1965)

Preface

Knowledge on the drying of sludge is comparatively limited, and publications thereon are scattered across journal articles. To date, no book devoted to belt drying of sludge has been written, and no comprehensive evaluation of the influence of sludge quality on belt drying and exhaust air treatment has been published. This book seeks to combine theoretical considerations with practical experience. The information it presents is intended to help interested parties in their understanding of the background of sludge drying and all related aspects.

To facilitate the understanding of technical expressions for Anglo-Saxon readers, the author has tried, as far as possible, to convert metric units into imperial units.

Deining, Germany
March 2022

Albert Heindl

Acknowledgments

The author would like to express his sincere thanks to HUBER SE, Berching, Germany, for providing photos, images, and information on sludge analytics, processing, and drying.

Other manufacturers of sludge belt dryers have likewise supported this book project with photos, images, and technical data, and the author would like to thank them for their assistance.

I would also like to thank my colleague, Dr. Mark Michaud, at the R&D department of HUBER SE, for fruitful discussions on the evaluation of GC-MS analysis of dewatered sludge samples and on the practical measurements of exhaust air from industrial belt dryers for sludge that he carried out over a number of years.

List of Symbols

C	Carbon in mg, g, or kg
CFU	Colony founding units, expresses in CFU per gramm of biofilter material or in CFU per mL of wastewater, corresponding to a number of vialble miorcirbial cells per g or mL
CHP	Combined Heat and Power Station
DM	Dry mass, expressed in kg
DR	Dry residue, which is used to describe sludge or biosolids already in a dewatered phase and determined by drying in an oven at 105 °C (221 °F). DR usually applies to dewatered and dried sludge, expressed in mass% relative to DM.
DR(A)	Maximum achievable Dry Residue DR with mechanical dewatering, eypressed in % and determined by slow drying, corresponding to the free water content of sludge. Procedure developped by Dr. Julia Kopp (Kopp 2001)
E	Elimination capacity in $g\,C/(m^3$ of a biofilter $\times$ h)
EPS	Extracellular polymeric substance
GPM	Gallons per minute = 3.78541 L/min
L	Liter, corresponding to a volume of 1 dm^3 = 0.264172 gallons
M	Molar mass weight of one mole of gas expressed in g/mole or kg/kmole or mass in kg
MGD	Million Gallons per Day 1 MGD = 157.725 m^3/h
NL	Normalized liter, a volume based on a temperature of 0 °C and a pressure of 101,325 Pa (standard volume)
P, PE	P refers to people's value, indicating capacity of a wastewater treatment plant. For example, 100,000 P means that the wastewater treatment plant is able to handle the wastewater of 100,000 people. People's values are expressed as the sum of the population whose wastewater is treated in a wastewater treatment plant plus additional people's equivalent value PE (calculated by BOD_5-freight from industrial wastewater divided by 60). For example: The production of vegetable or fruit conserves creates a PE value of 1 per 2 kg of vegetable conserves. Thus, in a town of 100,000 inhabitants, the P value of the wastewater treatment plant receiving the wastewater from a vegetable processing plant of 100 t per year is calculated to be 150,000.

PV	Protein value, characterizing the susceptibility of sludge to sticking, expressed as TKN×6.25/%DR in %
PS	Primary sludge, resulting from the mechanical treatment of wastewater
r	Back-mixing ratio
S	Surface in m^2
SBR	Sequenching Bed Reactor, akso named sequential batch reactor is a type of activated sludge process for the treatment of wastewater. SBR reactors treat wastewater such as sewage or output from anaerobic digesters or mechanical biological treatment facilities in batches.
t	Weight, corresponding to 1 mt = 1 metric ton = 1000 kg = 2204.6 lbs = 1.1025 imperial ton
T	Temperature in K
TOC	Total organic carbon
TKN	Total Kjeldahl nitrogen, expressed in kg N/kg DM
TS	Total solids, expressed in mg/L or as mass% of initial weight, which is used to describe a sludge that is still in a high liquid phase and determined by drying in an oven at 105 °C (221 °F). This means that dissolved substances, for example, salts, are considered.
TSS	Total suspended solids, expressed in mg/L or as mass% of initial weight, which is used to describe a sludge that is still in a high liquid phase and which is determined by filtering. This means that dissolved substances, for example, salts are not considered.
V	Volume in m^3
VDM	Volatile dry mass, expressed in kg
VDR	Volatile dry residue, expressed as organic dry mass relative to total dry mass in %, also referred to as loss on ignition
VOC	Volatile organic compounds, based on carbon, portion of total organic carbon (TOC) in sludge that is transferred to exhaust air during the drying process
VSS	Volatile suspended solids, which is used to describe sludge or biosolids still in a high liquid phase, determined by ignition in an oven at 550 °C (1022 °F), expressed in mass% relative to TSS
WAS	Waste activated sludge, sludge resulting from the treatment of biological sludge at a wastewater treatment plant
WWTP	Wastewater treatment plant
λ_C	Heat conductivity of sludge in $W/(m \times K)$
κ	Electrical conductivity in mS/cm
ϑ	Temperature in °C, with $°F = °C \times (9/5) + 32$
rec	Recovered

Contents

Introduction

1

"Biosolids are a product of the wastewater treatment process. During wastewater treatment, the liquids are separated from the solids. Those solids are then treated physically and chemically to produce a semisolid, nutrient-rich product known as biosolids. The terms 'biosolids' and 'sewage sludge' are often used interchangeably" (EPA 2022). Therefore, in the following chapters, the terms 'sewage sludge' or simply 'sludge' are used but also mean biosolids.

Sludge refers to any residual produced during primary, secondary, or advanced wastewater treatment without having been treated to reduce pathogens or vector attraction. The term sludge is further defined by process descriptors such as primary sludge (PS), secondary sludge, waste activated sludge (WAS) or biological sludge, aerated sludge (AS), and digested sludge (DS). Hydrolyzed sludge (HS) is characterized by thermic or thermochemical treatment of the sludge to rupture the cell structure of sludge particles and increase dewaterability.

To avoid pollution of ground water in the long-term, ever more countries are opting not to use sludge as a fertilizer. To recover at least the valuable phosphorus, a combination of drying and thermal oxidation (incineration) is useful. For instance, the extraction of phosphorus from the ash can be carried out by leaching with acid, adding other ingredients, and spray granulation. Drying sewage sludge in belt dryers is one important step to phosphorus recovery under the conditions of environmental protection with regard to the reduction of sludge transportation and the use of exhaust energy for drying.

This book not only deals with belt drying systems for sewage sludge to produce a dried and stable biological solid but also provides detailed information on the different types of sludge and their influence on drying behavior and exhaust air treatment. In recent years, a growing number of sludge drying centers have been planned and commissioned. One needs to bear in mind that sludge quality does not only vary from sludge to sludge due to the different treatment processes used and the varying ingredients in wastewater flux. Intermediate storage and the transport of dewatered sludge have a major impact on sludge rheology and on the specific sludge constituents transferred to the gas phase during the

A. Heindl, *Belt Drying of Sludge*, https://doi.org/10.1007/978-3-662-66512-1_1

drying process. There is also a seasonal effect: it is well known that winter sludge has a higher volatile matter content (fat, protein) and a lower achievable dry residue (DR) in dewatering than summer sludge. One needs to consider that microorganisms still exploit the organic mass in sludge cake, thereby changing the pH-value, DR, and rheology of the sludge, as well as generating methane and odorous gases with different odor thresholds, biodegradability, and water solubility.

Sludge drying is carried out since it has several advantages concerning further handling:

- Reduction in mass and volume
- Higher microbial stability in the case of full drying to a DR content of more than 87–90%
- Storage with minimum odor development
- Acceptable conveying properties, given that dried sludge is much like a common bulk material
- Increased calorific value
- Application as Class-A or -B fertilizer according to EPA 503 regulations, depending on drying system and temperature–time strain of sludge
- Use as a fuel for thermal processes such as incineration, pyrolysis and gasification

Thermal drying requires a certain amount of thermal and electric energy, which determine operating expenses (OPEX). For reasons of environmental protection as well as for economic reasons, waste energy—either from the wastewater treatment process itself or made available from external (industrial) sources—needs to be used as much as possible. In this context, belt drying systems are able to exploit exhaust energy sources with temperature levels starting at around 60–70 °C and reaching approximately 145 °C with regard to, for example, hot water. Lower temperatures require far higher belt drying areas and a considerable increase in airflow. Only in specific projects can very low waste energy temperatures be used under economic conditions. Higher temperatures are not used for safety reasons with regard to the minimum glow temperature of a 5 mm sludge dust layer of around 220 °C and the necessary distance to the process temperature of at least 75 K according to the European directive on potentially explosive atmospheres, ATEX.

Fig. 1.1 shows a two-belt dryer commonly used for sludge drying (Heindl 2021).

Sludge is a reservoir for harmful substances accumulating during the wastewater treatment process. One needs to bear in mind that these substances—heavy metals, pharmaceutical residues, enzymes, and other persistent substances from industrial processes—do not simply vanish, but remain in sludge or are partially transferred to the condensate of a condensation drying unit and/or to the exhaust air in general during all drying processes.

Fig. 1.1 Two-belt dryer, commonly used for sewage sludge drying. (Heindl 2021)

Before planning a belt drying system for sewage sludge drying, it is of the utmost importance to establish how dried material will be further used.

References

EPA (2022) Environmental Protection Agency, USA. https://www.epa.gov/biosolids/basic-information-about-biosolids. Accessed 29 Aug 2022

Heindl A (2021) Photo of a belt dryer installed at Emter GmbH, Altenstadt, Bavaria, Germany

2 Wastewater Treatment Plants and Processing Steps

Wastewater treatment results in the generation of sludge, the quantity of which—expressed in kg dry mass (DM) per inhabitant (P) and year (a)—strongly depends on the degree of stabilization. A rough estimation for sludge production can be indicated for:

- Non-stabilized sludge: $\approx$ 25–30 kg DM/(P × a)
- Aerobically stabilized sludge: $\approx$ 18–19 kg DM/(P × a)
- Anaerobically stabilized sludge (digested): $\approx$ 14–15 kg DM/(P × a)
- Thermally hydrolyzed sludge (with steam): $\approx$ 11–12 kg DM/(P × a)

The degree of sludge stabilization carried out varies considerably worldwide. It is influenced by the local possibilities for disposal or agricultural use, the specific costs per ton of dewatered sludge, as well as by official regulations. Thus, the type of stabilization ranges from non-stabilized sludge, which is treated and settled only mechanically and partially biologically, to full digestion with an integrated hydrolysis step.

The following list presents different wastewater treatments with an increasing degree of stabilization, while mechanical cleaning and settling is always performed when wastewater treatment is initiated:

- No further treatment (no stabilization)
- With a short period of aeration (poor stabilization)
- With a sufficient period of aeration of at least 15 days (aerobic stabilization)
- With digestion of at least 15–25 days depending on digester size (anaerobic stabilization)
- With hydrolysis and digestion (advanced anaerobic stabilization)

Consequently, sludge quality with regard to volatile dry residue (VDR), constituents such as protein, hydrocarbons, and fat as well as achievable dry residue (DR) after dewatering vary considerably. However, the treated water that is returned to nature must fulfill certain

A. Heindl, *Belt Drying of Sludge*, https://doi.org/10.1007/978-3-662-66512-1_2

thresholds for quality parameters such as chemical oxygen demand (COD) or nitrogen content, which are commonly linked to the size of the wastewater treatment plant.

One needs to bear in mind that all wastewater treatment steps have an effect on the quality of the dewatered sludge and, consequently, on the drying process itself. When changing parameters in processing steps, the impact on end-of-pipe operation, i.e., drying, must always be taken into account. For example, it is well known that biological phosphorus removal reduces the DR content of dewatered sludge by up to 4–5% DR (absolute). Thus, a higher amount of water needs to be evaporated in belt drying, and it is likely that the increased water-holding properties of dewatered sludge result in longer drying times and lower dryer throughput for a given dryer size. In addition to crystals of magnesium-ammonium-phosphates are formed and may cause wear in thick sludge pumps as well as conveying and extruding systems for dewatered sludge.

Fig. 2.1 below describes a wastewater treatment plant located in the town of Ingolstadt, Germany, with a wastewater treatment capacity of around 275,000 P. Table 2.1 describes all the technical features of this plant with digestion (ZKA Ingolstadt 2020). Fig. 2.1 presents the individual plant components. There are two belt dryers, each with three belts arranged one upon the other, that were commissioned in 2005 using exhaust energy from a nearby waste incineration plant at a very low exhaust temperature of 53–55 °C. Therefore, low temperature sludge drying is carried out, resulting in a higher specific electric and thermal energy demand per kg of evaporated water in comparison to medium-temperature belt drying. Nevertheless, this circular concept makes sense since it reduces sludge transportation requirements, and the calorific value of dried sludge is used to generate power for the waste incineration plant.

This example demonstrates how a belt drying system can be integrated into a closed mass and energy cycle in combination with a nearby waste incineration plant. The fact that the sludge drying plant has been in operation for more than 18 years (as of 2023) proves the economic and ecologic validity of this solution, even at very low-temperature exhaust gas energy.

Details of the wastewater treatment plant are presented in Table 2.1.

Fig. 2.2 illustrates a wastewater treatment plant in the United States (US) equipped with digestion and phosphorus recovery equipment, located in Howard County, Maryland (HDR Inc 2022). After the primary clarifier, sludge screening is carried out to remove tramp material before digestion. The digestion has a staged design for greater biogas generation. Biogas/natural gas is used for the direct heating of belt dryers. As sludge is heated to a temperature of at least 80 °C, dried granules can be used as a class A fertilizer according to US EPA 503 regulations. It can be used by local residents as garden fertilizer or to produce turf.

The simplified scheme of the plant is self-explanatory. For part of the dried sludge, a back-mixing step is integrated to make the dewatered sludge suitable for the extrusion process (see Chap. 5). As the DR of dewatered sludge is comparatively low for digested sludge, it can be assumed that there is a higher proportion of free water in dewatered sludge. Thus, sludge strains are more stable due to back-mixing and the increase in DR, and thixotropic flow of the sludge strains on the first drying belt is avoided. In Table 2.2 design data of this plant are mentioned.

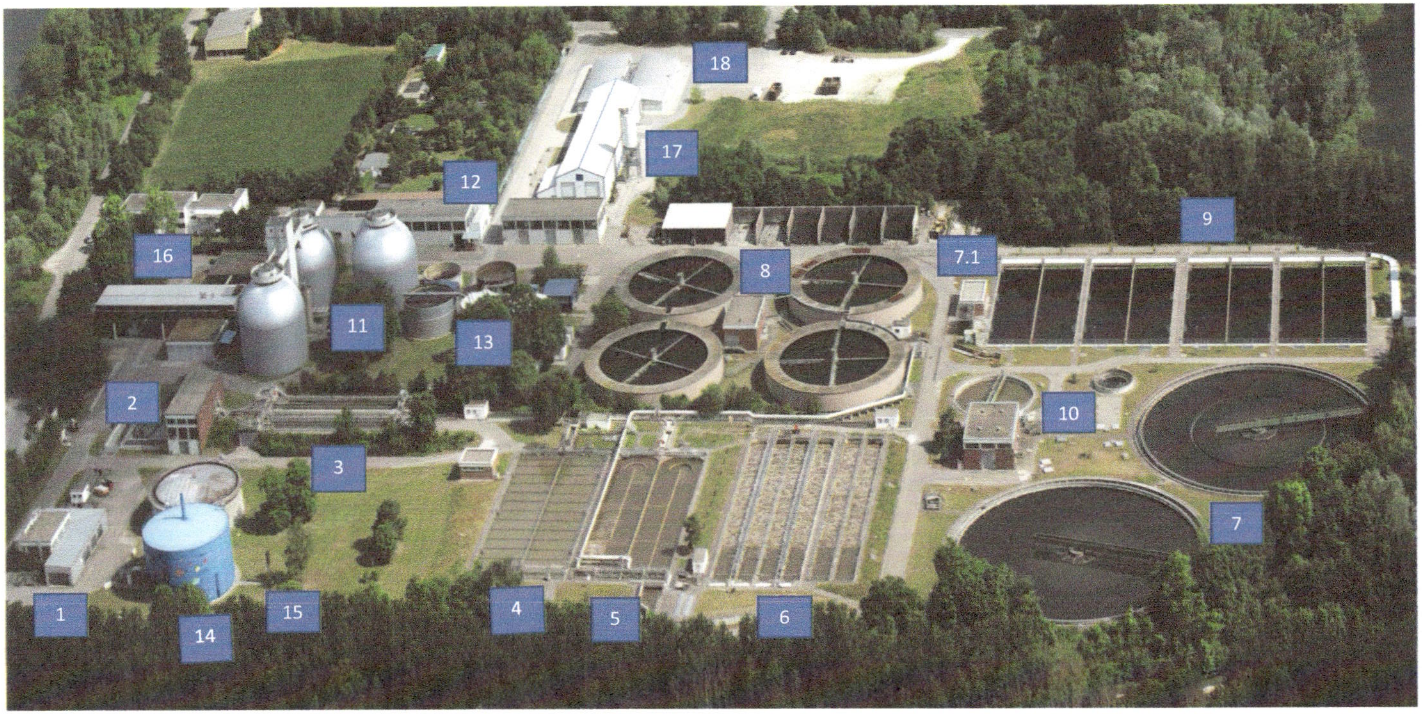

Fig. 2.1 Wastewater treatment plant for Ingolstadt, Germany. *1* Building for mechanical screen; *2* inlet pump station; *3* sand and grease trap; *4* primary settling tanks; *5* basins for denitrification and biological phosphate elimination; *6* aeration tanks; *7* intermediate treatment tanks; *7.1* facility for phosphate elimination with dosing of auxiliary agents; *8* trickling filters; *9* final treatment tanks; *10* sludge thickener; *11* digesters; *12* sludge dewatering facility; *13* process water treatment; *14* gas storage tank; *15* gas flare; *16* combined heat and power system (CHP); *17* belt drying lines; *18* exhaust air treatment (scrubbers and biofilters)

Table 2.1 Details of the wastewater treatment plant for Ingolstadt, Germany, (275,000 PE (people's equivalent)) with belt drying of sewage sludge. (ZKA Ingolstadt 2020)

	Data
Plant capacity	
Annual wastewater quantity	18,250,000 m^3/a
Dry weather effluent	60,000 m^3/day
Maximum dry weather effluent	3240 m^3/h
Mixed water discharge	6500 m^3/h
Biological oxygen demand over 5 days (BOD_5)	9400 kg/day
Total nitrogen (N_{total})	3015 kg/day
Total phosphorus (P_{total})	565 kg/day
Screen System	
Coarse screen with bars of 40-mm light distance	
Fine screen with bars of 6-mm light distance	
Step screen with 6-mm gap width	
Inlet pump station	
Screw pumps	2 × 1400 L/s and 1 × 850 L/s
Flood pumping station	
Propeller pumps	1 × 500 L/s, 2 × 1500 L/s
Sand separation	
Two aerated sand traps with grease separation chambers	2 × 1430 m^3
Primary settling tanks	
Volume	2 × 1430 m^3
Basins for denitrification and biological phosphorus elimination	
Volume	2 × 1308 m^3
First biological stage	
Volume	6 × 1485 m^3
Rotary blowers	4 × 4100 Nm^3 air/h
Sludge thickener	
Volume	1 × 1680 m^3
Intermediate settling tanks	
Two tanks each with a diameter of:	61 m
Volume	2 × 7070 m^3
Second biological stage	
Trickling filters	4 × 3400 m^3
Final settling tanks	
Volume	8 × 1128 m^3
Digesters	
Volume	3 × 4750 m^3
Sludge treatment	
Dewatering with centrifuges	3 × 20 m^3/h
TS of dewatered sludge	≈ 30% DR

Table 2.1 (Continued)

	Data
Sludge pumping by ≈ 50 m from centrifuges to 30 m^3 steel bins of each dryer by piston pump	
Dewatered sludge to two-belt dryer lines	≈ 2 × 0.9 m^3/h or ≈ 2 × 960 kg/h
	≈ 12,000 mt/a
Belt dryer lines (in operation for 16 years as of 2021)	2
Dried sludge	≈ 4,200 mt/a
DR of dried sludge	> 90%
Water evaporation	≈ 7800 mt/a
Hours of operation	≈ 8000–8100 h/a
Electric power demand	≈ 150 kWh/h or ≈ 0.154 kWh$_e$/kg of water evaporated
Thermal energy demand	≈ 7500 MWh/a or ≈ 1.04 kWh$_{th}$/kg of water evaporated
Exhaust air temperature	30–40 °C
Exhaust heat from condensing turbine of nearby waste incineration plant	≈ 53 °C/32 °C inlet/outlet of air heaters of dryers
Drying air temperature	≈ 46–48 °C
Ammonia content in uncleaned exhaust air	≈ 45 mg/m^3 at ≈ 35 °C
Odor units in uncleaned exhaust air	≈ 3000–6000 OU/m^3
Exhaust air	≈ 2 × 45,000 m^3/h
Volume of two biofilters	2 × 408 m^3
Odor units after scrubber and biofilter cleaning of exhaust air	< 500 OU/m^3
Benefits of drying and energy cycle with waste incineration plant	
Saving of truckloads for sludge transport	≈ 500/a
Saving of CO_2 emissions in truck transportation	≈ 176,000 kg CO_2/a
Gas utilization	
Biogas tank	4000 m^3
Gas engines (electric power)	1 × 220 kW$_e$, 2 × 370 kW$_e$
Exhaust thermal energy of gas engines	Used for sludge and digester heating as well as for heating of buildings
Self-generated electric energy by biogas incineration via CHP related to total electric energy demand of WWTP	≈ 56% (remark: electricity for dryer operation is supplied by the waste incineration plant nearby)
Process water treatment	
SBR	3 × 636 m^3
Flow of wastewater	Up to 450 m^3/d
Ammonium load	≈ 450 kg NH_4-N/d

DR, dry residue; *TS*, total solids; *CHP*, combined heat and power station; *WWTP*, wastewater treatment plant; *SBR*, sequenching bed reactor

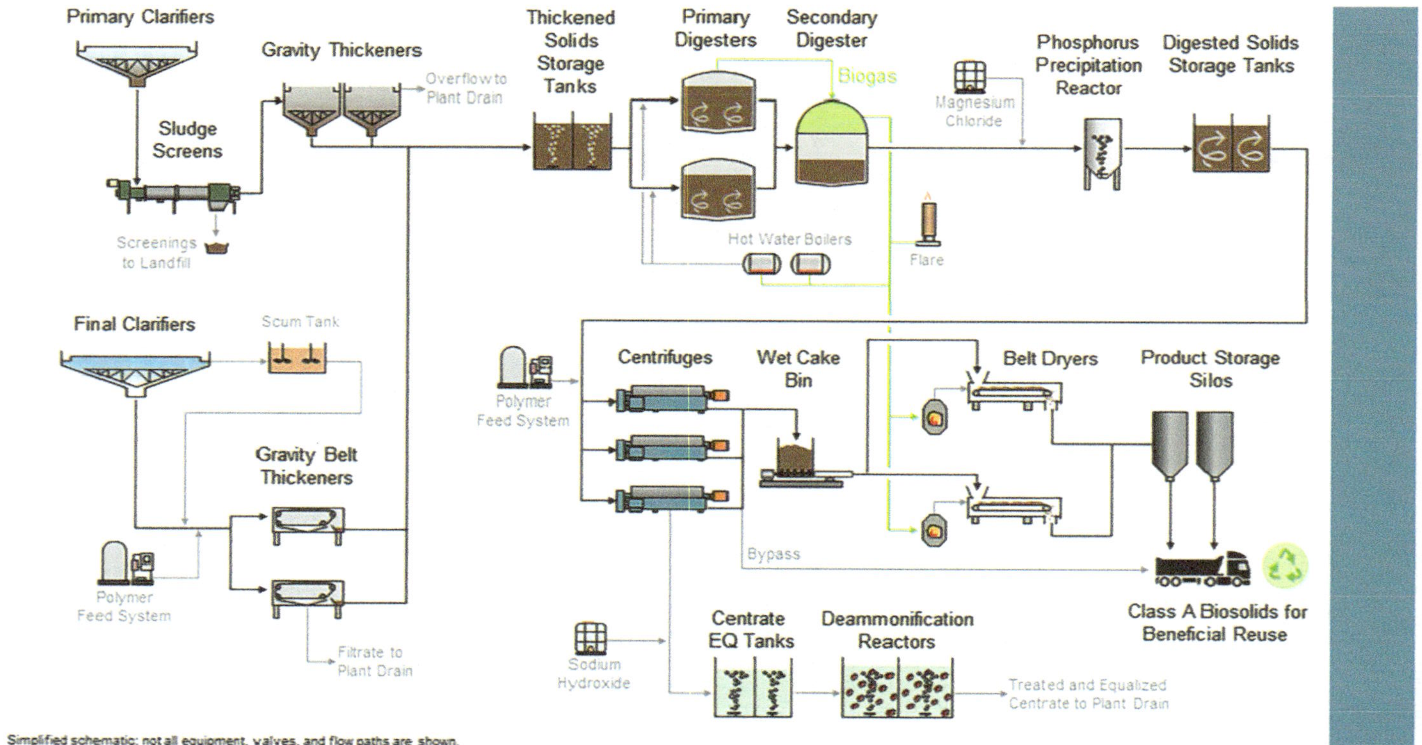

Fig. 2.2 Wastewater treatment plant in Howard County, US. (HDR Inc 2022)

Table 2.2 Technical data of the wastewater treatment plant in Howard County, US, with sludge digestion and belt drying (HDR Inc 2022)

Type of dryer	Two-belt dryer
Sludge preparation	With partial back-mixing of dried sludge to dewatered sludge
Heating	Directly fired with biogas/natural gas
Drying area	$2 \times 120\,m^2$
Dewatered sludge	$\approx 40{,}000\,t/a$
Feeding of dewatered sludge	$\approx 5000\,kg/h$
DR in/out	20/90%
DR of mixed sludge at dryer feed	≈ 30%
Water evaporation	$\approx 4000\,kg/h$
Specific water evaporation	$\approx 16.7\,kg/(m^2 \times h)$
Exhaust air treatment	With direct water spraying condensers

DR, dry residue

References

HDR Inc (2022) Engineering Firm. Information on Howard County project. Omaha, Nebraska. https://www.hdrinc.com/services/engineering

ZKA Ingolstadt (2020) Photo of the wastewater treatment plant for the town of Ingolstadt and details of a wastewater treatment plant in Bavaria, Germany

Types of Sludge and Their Properties 3

3.1 Types of Dewatered Sludge

3.1.1 Waste Activated Sludge or a Mixture of Primary Sludge and Surplus Biological Sludge

Primary sludge (PS) comes from the settling process and consists of easily decantable, large, and dense suspended particles. Its volatile suspended solids (VSS) content usually ranges from 55 to 60% relative to total solids (TS). Dewaterability is high and dry residue (DR) contents of 30–40% can be achieved.

Waste activated sludge (WAS) or biological sludge (BS) results from the biological treatment of wastewater and mainly consists of microorganisms, i.e., bacteria and other volatile compounds. WAS is characterized by low stabilization or no stabilization at all. VSS range from 80 to 85%, and in extreme cases up to 87%. The content of total suspended solids (TSS) usually ranges from 7 to 10 g/L. Dewaterability is fair to medium and DR values of 14–17% after dewatering are possible. For mixed sludge, i.e., a mixture of primary sludge and WAS, the VSS values range from 75 to 80%. WAS is sometimes aerated for a few days in storage tanks, thus improving the degree of stabilization and reducing VSS to 70–75%. Aeration is carried out with 3.3–3.8 kg air/kg of volatile dry mass (VDM) for 7–9 days, and a reduction in volatiles of up to 40% is achievable. During cold digestion without aeration at 10 °C for 10 days, the degradation ratio for volatiles is up to 5% relative to initial VDM. This value is calculated with the data from Fair and Moore (1937) assuming 50% efficiency.

Fig. 3.1 shows a mixed sludge of primary sludge and biological sludge (WAS). Their mass fraction ranges from 40 to 50% for PS and from 50 to 60% for WAS. Total Kjeldahl nitrogen (TKN) contents range from 50,000 to 90,000 mg/kg of dry mass (DM). Ammonium (NH_4–N) contents cover values from 5000 to 18,000 mg/kg DM.

High protein contents are also associated with a high bacterial content. *Escherichia coli*, for example, have the following composition as shown in Table 3.1 (Kunz 1992).

A. Heindl, *Belt Drying of Sludge*, https://doi.org/10.1007/978-3-662-66512-1_3

Fig. 3.1 Mixed sludge formed from primary sludge and waste activated sludge. (HUBER SE 2018a)

Thus, the volatile content (volatile dry residue [VDR]) of bacteria DM is 93.5% and the protein content is 50%. As a consequence, the volatile mass of WAS with a high portion of bacteria will always have higher levels of protein.

The DR of dewatered sludge often ranges from 15 to 20%; lower values are common especially when sludge is dewatered by low-efficiency belt presses and no special tests for polymer optimization have been undertaken.

The electrical conductivity of liquid WAS or mixed PS/WAS sludge does not exceed 1 mS/cm as a rule, and the DR of mixed sludge achieved by dewatering is typically below 22%, even with high-efficiency dewatering equipment. The pH-values of WAS normally range from 5.5 to 6.5, resulting in a high load of sulfuric components in exhaust air from belt dryers.

Microorganisms show high activity, and organic matter is decomposed rapidly in stored dewatered sludge, reducing the pH-value even further in sludge bulk. The DR content is reduced due to hydrolysis and structural strength of sludge is decreased consequently. Dewatered WAS exhibits marked colloid behavior, especially when a higher portion of

Table 3.1 Composition of *Escherichia coli*. (Kunz 1992)

Substance	Mass% relative to original substance	Mass% relative to dry mass
Water	80	400
Proteins	10	50
Sugar	2	10
Fat	2	10
Amino/volatile acids	1.3	6.5
Desoxyribonuclein Acid (DNA)	0.4	2
Ribonucleic acid (RNA)	3	15
Non-volatiles	1.3	6.5

wastewater from breweries and dairy industries is processed. Deposits of fatty or oily substances in condensing units can often be observed, which can result in blocking and frequent cleaning of heat exchangers.

Extrusion of dewatered sludge is possible even at very low DR of 15%, but the stability of formed strains is lower in comparison to digested sludge or sludge stabilized by aeration. During drying, sticking of extruded strains and plate forming in the material bed occurs in the case of higher protein values. Sticking behavior is enhanced by higher sludge temperatures and denaturation of temperature-sensible proteins. The shear stress of sludge as a measure for sticking intensity increases with augmenting DR and typically reaches its maximum at a DR of 40% (KEMIRA 2020). As a result, plate forming might occur in belt drying of WAS or a PS/WAS mixture.

Successful belt drying of WAS or a mixture of primary and secondary sludge is linked to the individual processing steps:

- If possible, further destruction of volatiles by proper aeration of thin sludge before the dewatering step.
- Selection of the correct cationic polymer to optimize dewatering with regard to DR.
- pH adjustment, for example, by adding NaOH (increasing pH) or $FeCl_3$ (lowering pH), influences odor emission as the transfer of odorous gas components is controlled or avoided. The effect of pH on polymer efficiency is negligible in most cases.
- The addition of polyaluminum chloride (PACl) to thin sludge reduces sticking behavior of dewatered sludge during drying. However, in the case of the pH-value decreasing to below 5, floc formation and stability is negatively affected and the pH-value needs to be increased. For example, a PAX-14 ($Al_2Cl(OH)_5$, 174.45 g/mole, Al by weight $7.2 \pm 0.3\%$, 3.5 ± 0.1 mol/L, pH < 1, ligand number r = 0.8, basicity B = 26 ± 6%) solution of 10 L is added to 5 m^3 of thin sludge and sludge flocs are covered with clusters of PACl, thus shielding the underlying sticky biopolymers when the range of DR with sticking behavior of sludge is surpassed, as shown in Fig. 3.2 (KEMIRA 2020; Peeters 2011). PACl always needs to be mixed into sludge with sufficient turbulence.
- Minimizing storage time between dewatering and drying to reduce sludge degradation with regard to structure and decomposition of organic matter.

Aeration tanks or basins for WAS always need to be properly operated in order to function efficiently for a minimum of stabilization and to limit odor issues (WEF 1995):

- Regular inspection and cleaning of air diffusors
- Even mixing of activated sludge and avoidance of solids depositions in corners or edges
- Minimum air flow of 0.3 m^3 per meter of tank length and per minute
- Diffused air systems with a minimum of 20–30 m^3 of air per 1000 m^3 of tank volume and per minute
- Power concentration for mixing with mechanical aerators of 13–26 kW per 1000 m^3 of tank volume

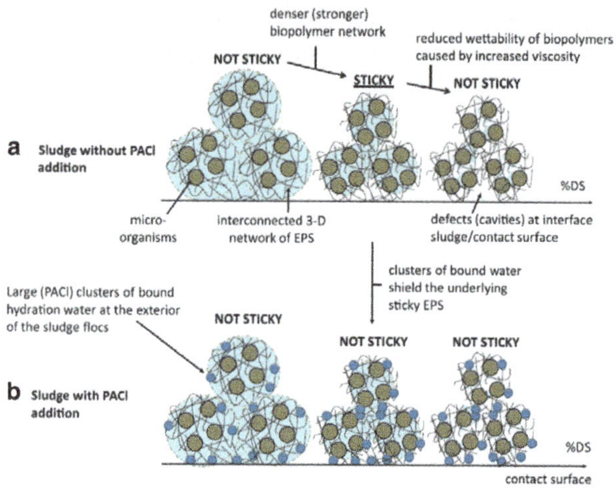

Fig. 3.2 Effect of adding polyaluminum chloride (*PACl*) to waste activated sludge or a mixture of primary and secondary sludge to prevent sticking during drying. (KEMIRA 2020). *EPS*, extracellular polymeric substances

Belt drying of WAS or a mixture of primary and biological (secondary) sludge is always related to a loss in specific water evaporation in comparison to digested sludge due to the fact that case hardening and sticking occurs during drying, and drying time is considerably longer.

The predominant odorous gas components in exhaust air leaving belt dryers are sulfur-based, such as hydrogen sulfide (H_2S), dimethyl sulfide [$(CH_3)_2S$], dimethyl disulfide [$(CH_3)_2S_2$], or methane thiol (CH_4S) as the pH-value of sludge ranges from 5.5 to 6.5. Ammonium (NH_3)-concentrations are far lower in comparison to the values commonly encountered for anaerobically stabilized sludge. An exhaust air measurement carried out in the USA (southeastern region) for a belt dryer with a condensing system resulted in the exhaust air gas concentrations shown in Table 3.2. The pH-value of sludge was comparatively low at 5.6. Thus, the values for sulfur-based gas components were quite high (factor of 10 higher than in comparison to digested sludge), whereas the ammonium value is quite low (more than a factor of 10 less than in the case of digested sludge). The measurements were made downstream of the condensing unit at an exhaust air temperature of around 43 °C and a relative gas humidity of around 100%. Sludge fed to the dryer had a DR of 21–23%

Table 3.2 Gas components in exhaust air of a belt dryer drying a mixture of primary sludge and waste activated sludge. (HUBER US 2018)

Gas component	Concentration in exhaust air in mg/m^3	Estimation of specific gas emission in mg/kg dry mass of processed sludge
Ammonia (NH_3)	≈ 15.7	≈ 330
Hydrogen sulfide (H_2S)	≈ 23.6	≈ 490
Methane thiol (CH_4S)	≈ 28.3	≈ 590
Dimethyl sulfide [$(CH_3)_2S$]	≈ 3.6	≈ 75
Dimethyl disulfide [$(CH_3)_2S_2$]	≈ 5.8	≈ 120

and a VDR of approximately 80%, while air temperature was 95 °C. As the mass ratio of dimethyl sulfide and dimethyl disulfide to methane thiol are far less than 1, it can be concluded that the sludge was dried immediately after dewatering and no additional, prolonged intermediate storage or oxygen transfer to sludge due to mechanical handling occurred.

3.1.2 Aerated Sludge

Stabilization is carried out by intense aeration of sludge for at least 15 days. The average VDR of dewatered sludge is around 68% relative to DM, while VDR ranges from 65 to 75%. Dewatering reaches 21–25% DR; however, lower values of DR and higher values of VDR might be encountered when aeration time is shortened to save electric energy in wastewater treatment. The density of sludge was determined to be 1070–1080 kg/m^3. The electrical conductivity of aerated thin sludge covers a broad range of 0.3–9.8 mS/cm and averages 1.7 mS/cm (Deutsche Vereinigung für Wasserwirtschaft, Abwasser und Abfall e.V., DWA 2008). In central Europe, specific electric energy consumption for aeration ranges from 27 to 32 kWh/(P × a) (Kroiss and Svardal 2009).

Fig. 3.3 shows dewatered sludge stabilized by aeration. The sludge has a brown color and an earthy smell. As a rule, protein values (TKN × 6.25/%DR) are comparatively low (18–30%), and no sticking problems occur during belt drying. In Germany, aeration stabilization is carried out in smaller wastewater treatment plants with less than 15,000 P. In the last decade, a trend has developed to save electric energy by reducing aeration time. At first glance, this might be positive for environmental protection, but with a reduced degree of stabilization, higher emissions of methane need to be accepted during storage of dewatered sludge. The oxygen concentration in the aeration tank of municipal wastewater plants needs to be kept in the region of 1.0–2.5 mg O_2/L (NRW 2018), thus fixing aeration flow of air. Lowering the O_2 concentration to below 1.0 mg/L may result in N_2O formation due to oxygen stress exerted on microorganisms and formation of swimming sludge on top of the water level of the aeration tank.

The degree of stabilization is determined by measuring the oxygen demand, a procedure that is described in Chap. 14. Well-aerated dewatered sludge (> 15–20 days of aeration with sufficient oxygen concentration) is highly suitable for extrusion and belt drying.

3.1.3 Digested Sludge

Digested sludge results from a mixture of primary and secondary (biological sludge) that has been treated under anaerobic conditions for several weeks in a digester at a temperature of around 34–38 °C. Properly performed digestion produces a well-stabilized sludge with high degradation of organic dry matter, often also named volatile dry matter (VDM). The technical limit of VDM degradation ranges from 50 to 70%, depending on the constituents of the sludge VDM. In anaerobic digestion, carbohydrates, proteins, and fats are

Fig. 3.3 Sample of dewatered sludge stabilized by aeration. (HUBER SE 2017)

degraded to a biogas, mainly consisting of methane (60–70 vol%) and carbon dioxide (30–40 vol%). As a rule, sludge generation in Central European municipal wastewater treatments plants is around 14–15 kg DM/(P × a). The electrical conductivity of digested thin sludge ranges from 1.1 to 15.2 mS/cm. The average pH-value of thin sludge after digestion is around 7.2–7.5 (DWA 2008). Dewatered sludge usually reaches pH-values above 8.

The process of digestion consists of several steps (Fig. 3.4):

- *Hydrolysis of proteins* by enzymes, forming peptides and amino acids being the origin of hydrogen sulfide and ammonia gases, hydrolysis of fats forming glycerin and fatty acids, hydrolysis of carbohydrates forming mono- and disaccharides (typical pH-range 5.2–6.3), exo-enzymes degrading solid chemical oxygen demand (COD) into dissolved COD
- *Acidogenesis* forming acetate, ethanol, lactate, butyrate, propionate as well as hydrogen and carbon dioxide
- *Acetogenesis* (typical pH-range 6.7–7.5), degrading the abovementioned products to acetate
- *Methanogenesis* (requiring a pH-value of 6.7–7.5), degrading acetate to methane and carbon dioxide

In the digestion of municipal sludge with limited industrial wastewater intake < 15–20 mass%, VDR of sludge averages 58% with a range of 52–62%. In the case of a high input of co-substrates from the food industry, higher VDRs of up to 68% are encountered. Normal municipal digested sludge can be dewatered to a DR of 26–30%, provided that polymer selection and dosing are properly performed. However, large wastewater treatment plants with mono-incineration of partially dried sludge dewater to a DR of 22–24%

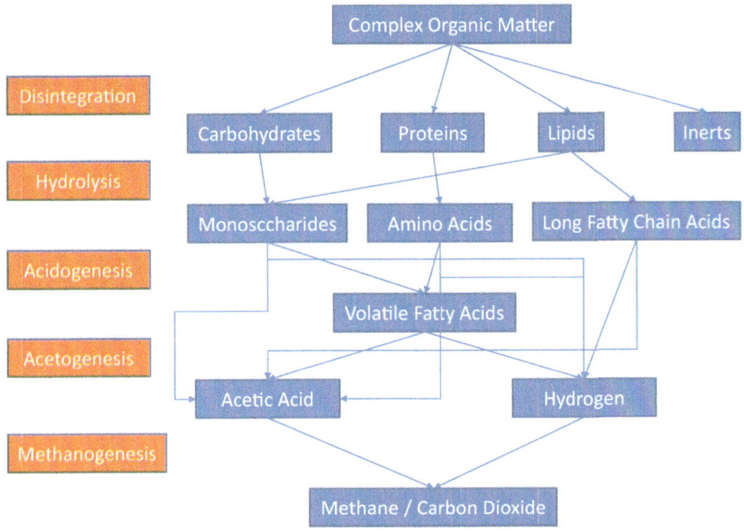

Fig. 3.4 Degradation steps of volatile dry mass during anaerobic digestion. (According to Panico et al. 2014)

to save polymer costs and accept higher thermal energy consumption at lower specific costs per kWh_{th}. Also, a high input of co-substrates from the food industry reduces the achievable DR of dewatered sludge.

Dewatered digested sludge has a black appearance due to its higher content of iron sulfide (FeS; Fig. 3.5). During digestion, H_2S reacts with Fe ions to form FeS. Its odor is earthy with a slight impression of tar. Sufficiently dewatered sludge is crumbly when leaving the dewatering machine.

Fig. 3.5 Digested dewatered sludge. (HUBER SE 2016)

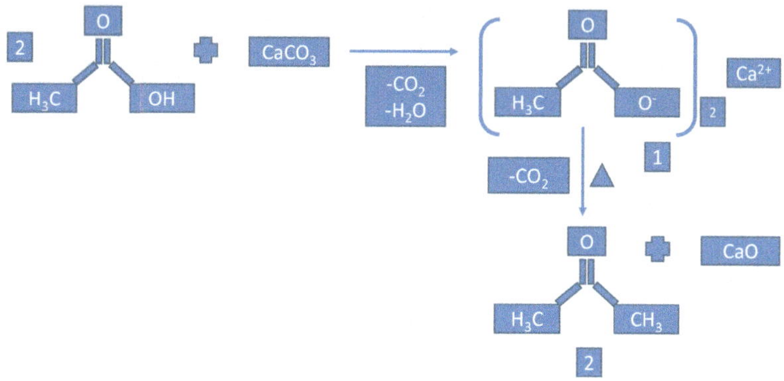

Fig. 3.6 Formation of acetone in the digestion process. (According to Wikipedia 2022)

The protein value is low to medium and there is no or only a slight sticking effect during belt drying of properly digested sludge.

The pH-values of the dewatered sludge range from 7.8 to 8.6 immediately after dewatering. There is an increase in pH of up to 0.5–1.0 since CO_2 is stripped out of sludge during the dewatering process. The rise in pH from thin sludge to dewatered sludge increases with the physical height of digesters and the additional mass of CO_2 stored in sludge under pressure.

The mass-dominating odor gas component is ammonia (NH_3), but high contents of inorganic and organic sulfur components such as hydrogen sulfide (H_2S) and methane thiol (CH_4S) may increase the odor contribution in exhaust air even at high pH-values. Especially dewatered sludge with longer intermediate storage ($\gg 3$ days) suffers from a decline in pH-value and the additional formation of carbon acids resulting from ongoing biological degradation of VDM.

If the anaerobic digestion process is insufficient, intermediate compounds such as volatile acids and acetates will remain in the digested sludge when leaving the digester. After dewatering of the sludge and its transfer to the dryer, volatile compounds of this kind end up in the exhaust air of the dryer. Such gas components and their further reaction compounds may cause problems in the exhaust air treatment system. For example, propionic and acetic acid further increase the load on alkaline scrubbers.

Calcium acetate ($C_4H_6CaO_4$) decomposes when the final sludge temperature exceeds 60 °C, and acetone (C_3H_6O) is generated during the drying process according to Fig. 3.6. Acetone not separated in the scrubbers additionally burdens the components of downstream exhaust air treatment.

Improper digestion due to excessive co-substrate or industrial wastewater intake from the food industry—especially with peak loads—, hydraulic overloading of the digester, or insufficient thin sludge circulation/mixing significantly increases the fat and protein content of digested dewatered sludge and increases the risk of case hardening during belt drying. A barrier of dried protein and fat layers is formed at the surface of sludge granulates and increases drying time. As a direct consequence, dryer throughput might be significantly lowered.

3.1.4 Sludge from Wastewater Treatment with Biological Phosphorus Elimination

Some larger wastewater treatment plants in Germany use biological phosphorus (Bio-P) elimination. By applying alternating aerobic and anaerobic conditions, special microorganisms are forced to deposit phosphorus as insoluble polyphosphates in their cells. These microorganisms are separated with biological (secondary) sludge. The advantages of this process include a remarkable saving of precipitation agents, for example, $FeCl_3$, and a reduction in inorganic dry mass of sludge. However, in comparison to phosphate precipitation with iron or aluminum salts, there are also some disadvantages. The Bio-P process results in a higher orthophosphate back-loading to the biological process step of the wastewater treatment plant, in worse sludge qualities with regard to the thickening and dewatering process, and in a surplus demand of flocculating agents of up to 3 g of efficient substance per kg of dry mass of sludge (Kopp 2017).

In the case of digested sludge, the absolute loss of DR of dewatered sludge ranges up to 3–5% in comparison to digested sludge with chemical phosphate precipitation (Meinel and Gnirss 2014; DWA 2008). However, the addition of chemical agents increases DM and there is certainly a proportion of apparent gain in DR for chemical precipitation. For aerobically stabilized secondary sludge, the DR of dewatered sludge declines from approximately 18 to 14–16%, or by about 2 to 4% DR absolute (Higgins 2016).

In addition, there is a problematic issue in the operation of wastewater treatment plants with regard to the formation of magnesium ammonium phosphate crystals $[(NH_4)MgPO_4 \times 6\,H_2O]$—also called struvite—by back-solving of phosphates under anaerobic conditions.

$$Mg_2{}^{+} + NH_4{}^{+} + HPO_4{}^{-} + 6\,H_2O \Leftrightarrow MgNH_4PO_4 \times 6\,H_2O + H^{+} \tag{3.1}$$

Uncontrolled precipitation of struvite crystals and deposits occur in pipes for the transportation of digested thin sludge as well as in pumps, tanks, basins, and dewatering equipment, due to the stripping of CO_2 and to an increase in the pH-value of sludge (Heinzmann and Lengemann 2014). Stripping of CO_2 during the dewatering of sludge increases the pH of dewatered sludge, thereby promoting struvite formation. The minimum of solubility of struvite is in the region of pH = 8–11. Struvite precipitation increases with augmenting K_{SP} value, formed by multiplying concentrations of $c(Mg_2^{+}) \times c(NH_4^{+}) \times c(PO_4^{3-})$—each

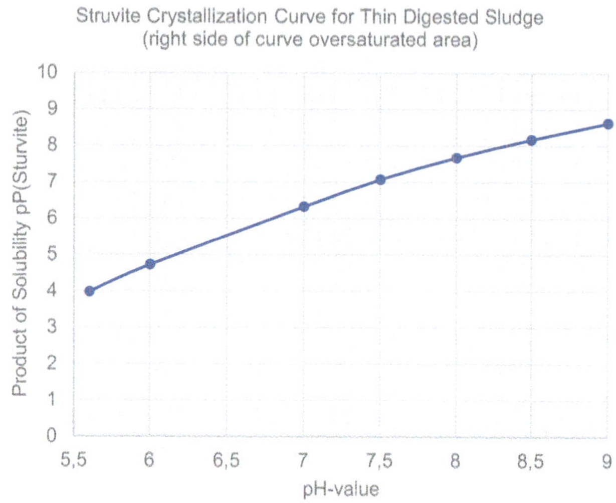

Fig. 3.7 Struvite crystallization curve

in kg/L—and with rising temperature (Ramadhania 2021). The saturation curve for struvite is calculated by the product of solubility p_P (struvite):

$$p_P = -\log K_{SP} = -\log\left[c(Mg^{2+}) \times c(NH_4^+) \times c(PO_4^{3-})\right] \tag{3.2}$$

Fig. 3.7 illustrates the curve of struvite crystallization, where precipitation of struvite happens on the right side of the curve. In the case of a pH-value of, for example, 7.5 of digested thin sludge, precipitation of struvite will occur as soon as the product of solubility p_P is below 7.1 (Ohlinger et al. 1998)

The curve can be described by the following formula (3.3) (Fig. 3.7):

$$p_P = 0.1677 \times pH^2 + 3.8171 \times pH - 12.12 \tag{3.3}$$

With $R^2 = 0.9999$.

The PO_4–P content in thin sludge after digestion is around 300 mg/L for wastewater treatment plants with Bio-P (Baudirektion Kanton Zürich 2013); however, maximum values of up to 500 mg/L might occur. The content of orthophosphate generally ranges from 50 to 200 mg/L (BAFU 2009), with an average value of around 125 mg/L, yielding 40% of the average total PO_4–P content. If one takes 150 mg orthophosphate/L, a struvite content of 517 mg/L can be calculated. At a total solids (TS) of 2.7% in thin sludge (27 g/L), the struvite concentration is around 1.44 mass% relative to TS. Neglecting TS losses during dewatering, such concentrations can also occur in dewatered sludge. However, the pH-value in dewatered sludge will rise due to stripping of CO_2 during dewatering, causing increased depreciation of struvite inside belt dryers.

It must be taken into account that, depending on grain size and content of struvite crystals in digested and dewatered sludge, higher wear and deposits on conveying pumps as

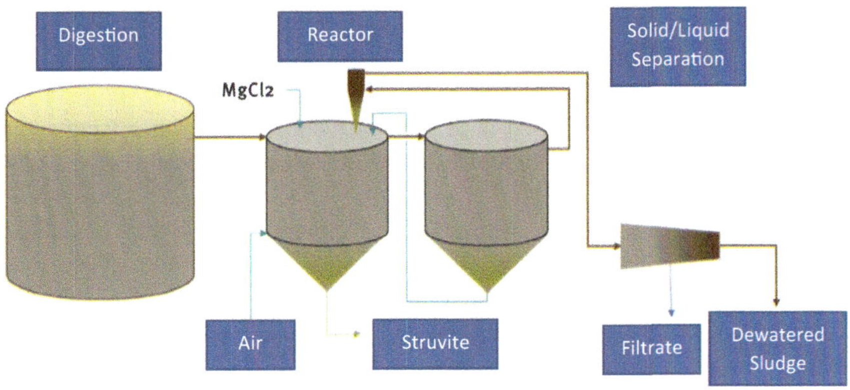

Fig. 3.8 Unit for the removal of struvite crystals according to AirPrex® Macro process. (CNP CYCLES 2021)

well as on machines for the mechanical transport of dewatered sludge may occur. Dewatering equipment such as centrifuges or screw presses are liable to abrasion inside and at discharge positions. In the case of belt dryers, intense surface wear in extrusion units, conveying units, and mixing units for fully dried and dewatered sludge might occur. Using materials with higher wear resistance solves the problem but increases expenditure.

Practical experience shows that for struvite concentrations of more than one mass% in digested sludge (relative to DM), increasing wear on sludge treatment equipment needs to be accepted. For this reason, struvite removal needs to be considered to limit wear in downstream wastewater treatment steps.

To limit uncontrolled and spontaneous precipitation of struvite, there is the possibility of stripping carbon dioxide (CO_2) under controlled conditions. The pH-value is increased from 7.7 to 8.4 and magnesium chloride ($MgCl_2$) is added to help the precipitation of struvite crystals out of thin sludge. These crystals are discharged from the main flow of digested thin sludge. The pipeline between the digester and the processing step can be kept free from deposits by adding inhibitors that prevent solidification. The efficiency of struvite removal can be enhanced by a two-stage arrangement and back-mixing of smaller crystals to the main reactor (CNP CYCLES 2021). Fig. 3.8 shows a treatment unit of this kind for the removal of struvite after digestion.

The efficiency of precipitation and removal of struvite crystals depends on the following factors:

- The pH-value needs to be set in a slightly alkaline range by stripping of CO_2 and, if necessary, by adding of NaOH.

- The presence of calcium, aluminum, and iron ions, competing with Mg ions, affects the formation of magnesium phosphates. Low levels of these substances promote struvite formation and the removal of crystals.
- The feeding of $MgCl_2$ is carried out hyperstoichiometrically with a factor of $\beta =$ 1.3–2.5. Thus, around 1.3–2.5 kg $MgCl_2$ per kg PO_4 are used.

The orthophosphate, stored in digested sludge is precipitated by 90–95%. The efficiency of struvite crystals removal reaches 30–40%, and grain sizes of over 500 µm are removed from sludge without the use of a hydrocyclone. An analysis of grain size distribution showed that 60 mass% of crystals have a size of below 500 µm, while the residual mass% corresponds to a grain diameter of 500–1500 µm. Using a hydrocyclone with a separation threshold based on grain size of approximately 50 µm leads to higher removal, especially of smaller crystals.

The specific electric energy demand without a hydrocyclone ranges from 1.1 to 1.3 kWh/m^3 of thin sludge, and with a hydrocyclone from 2.5 to 3.5 kWh/m^3 (CNP CYCLES 2021).

Thus, the orthophosphate content in cleaned wastewater at the exit of the wastewater treatment plant can be reduced to a relative value of 5–10% of feed concentration in wastewater fed to the plant.

The removed struvite crystals, consisting of approximately 62 mass% pure struvite, 10% water, 9% organic matter, 10% sand, and 9% other minerals, can be used as fertilizer according to the German regulations on fertilizers (Heinzmann and Lengemann 2014). Thus, the recovery of phosphorus from sludge is made possible, reaching around 5–20 mass% of phosphorus fed to the wastewater treatment plant.

3.1.5 Hydrolyzed Sludge

There are several systems for the thermal and thermochemical hydrolysis of sludge. Hydrolyzed sludge (HS) comes from a hydrolysis process to rupture cells of sludge microorganisms, either carried out by steam heating (140–165 °C approximately 6–7 bar[a]) for 20–30 min and rapid steam pressure drop or by alkaline treatment at a moderate temperature of 60–70 °C for 120 min. In the past, the most common arrangement of hydrolysis was upstream of digestion, but it is also possible to place the process between two digesters, to treat only the WAS and bypass primary sludge, or to set up a post-digestion configuration.

The TSS content of hydrolyzed sludge may typically reach 80–120 g/L before final dewatering; in some cases, up to 200 g/L is possible. The DR content of dewatered hydrolyzed sludge generally ranges from 30 to 45%.

The best known hydrolysis process is the CAMBI thermal hydrolysis process (THP), which works with steam at 165 °C and 6 bar(a). The cell walls of treated sludge are ruptured and a higher portion of VDM is available for digestion. The main goals and features of this special sludge treatment using steam for the usual arrangement before the digestion process can be stipulated:

- Reduction of sludge mass due to higher degradation of VDM in sludge during digestion (values of up to 60% of VDM degradation are possible with the CAMBI process, with typical values being 56–58%).
- Higher dewaterability of sludge (+5 to +9% absolute in DR are possible in comparison to conventional digestion); DR values of up to 45% have been achieved in the past, with typical values being between 30 and 40%.
- Decrease in polymer consumption for dewatering.
- Higher biogas generation (+13 to +25% in comparison to conventional digestion).
- High level of hygienization of sludge prepared for use as a fertilizer.
- The steam requirement for THP is around 600–1000 kg/mt of DM fed to system, depending on degree of steam reuse and system arrangement. Specific electricity demand ranges around 0.14 kWh/kg of DM fed to system (Abu-Orf and Goss 2012).

Sludge treatment can be carried out either batch-wise or continuously and by heating/expanding with vapor steam or by temperature/alkaline processing. A higher odor load related to higher contents of volatile acids and sulfuric volatile organic compounds (VOC) needs to be considered for the ventilation of sludge transportation units after the hydrolysis process and in the exhaust air of belt dryers. Special dewatering equipment such as chamber filter presses or hydraulic filter presses are used to achieve the highest possible DR in dewatering. Drying time is considerably reduced by 50–67% of the value of conventionally digested and dewatered sludge, since less water needs to be removed and the water-releasing ability of hydrolyzed sludge is much higher on the basis of ruptured cells. This means that a smaller belt dryer can be used and energy requirements (thermal and electric) will be considerably reduced. The effect of the thermal hydrolysis process depends on sludge characteristics. The highest efficiency of hydrolysis and subsequent dewatering can be achieved with WAS, whose water-holding capacity is highest under normal wastewater treatment conditions. Pre-digestion of hydrolyzed WAS is carried out, and the treated WAS is mixed with PS for final conventional digestion (CAMBI 2022).

Fig. 3.9 shows a CAMBI hydrolysis unit installed at the Crossness wastewater treatment plant/Thames Water, UK (HUBER SE 2019)

3.1.6 Industrial Sludge from Pulp and Paper Mills

Industrial sludge often differs considerably from municipal sludge. Therefore, individual analyses need to be carried out and behavior in extruding and drying needs to be examined individually. Some experience has been gained with sludge from the waste of pulp and paper mills produced by biological wastewater treatment. Fiber content is an important issue with regard to dewatering, extruding, and drying. For reasons of extrusion and

Fig. 3.9 CAMBI sludge processing unit at the Crossness wastewater treatment plant operated by Thames Water, UK. (Heindl 2016)

plate-forming during the drying process, the fiber content needs to be limited to less than 20 mass% of DM.

Sludge that has been successfully belt dried was a mixture of secondary biological sludge and tertiary precipitation sludge. The precipitation agent was $Al_2(SO_4)_3$. The degree of dewatering needs to be set to the highest possible value, at least not less than 2% absolute below the maximum possible achievable DR (named DR(A)) of sludge, determined by a controlled drying process in the laboratory. Fiber content needs to be less than 10% relative to DR; restricted fluctuations of content are ±10% relative. High fiber contents might block extrusion units or mixing devices when fully dried sludge is mixed with dewatered sludge to a DR of 65%. The xylene content in fiber sludge needs to be less than 12% relative to DM if the sulfate process of paper manufacture is used. The volatile content (VDR), determined at 550 °C, must not exceed 70–75% relative to DM; higher values of secondary biological sludge (e.g., up to 86% DR) might result in higher VDR of the sludge mixture and consequently in plate formation or sticking in the dryer. Fig. 3.10 shows dried sludge at the discharge point of the second belt of a two-belt-dryer (HUBER SE 2019).

The susceptibility of the sludge mixture to sticking and plate formation during belt drying is caused by contents of the hemicellulose xylene and fibers. This effect might reduce dryer evaporation and consequently dryer throughput. The sticking effect can be significantly reduced by adding polyaluminum chloride (PACl) before dewatering, as shown in laboratory tests. Its application needs to be individually adapted to the type of sludge for each project.

Waste incineration is often carried out at pulp and paper mills, thereby providing the opportunity to further use the energy content of exhaust gases for the drying of paper sludge.

Fig. 3.10 Dried sludge from a pulp and paper mill leaving the dryer at the discharge point of the second belt. (HUBER SE 2019)

3.2 Types and Degree of Sludge Stabilization

There is a broad spectrum of types of municipal sludge, mainly influenced by wastewater intake, the process steps of wastewater treatment, and consequently the degree of stabilization. The following types of sludge may be encountered with minor variation in processing steps during wastewater treatment:

- Non-stabilized, biological sludge (waste activated sludge [WAS]) or mixtures of primary sludge (PS) and biological sludge (BS).
- Aerated sludge, aerobic treatment with intense air intake for degradation of volatile matter in sludge. Aeration is normally done at smaller wastewater treatment plants < 15,000–20,000 P.
- Digested sludge, anaerobic treatment with exclusion of oxygen, and formation of methane and carbon dioxide out of volatile matter. Degradation of volatile matter can be increased by using thermal or thermal/chemical hydrolysis.

The higher the degree of sludge stabilization, the higher the DR of dewatered sludge and the lower the capital expenditure (CAPEX) and operating expenses (OPEX) for a belt drying unit. There is a harsh saying concerning sludge stabilization on wastewater treatment plants:

> *"Everybody gets the (dewatered) sludge he deserves!" This means that low effort and cost for sludge stabilization in a wastewater treatment process always result in high demands in terms of investment and operating costs for sludge drying.*

Table 3.3 Ratio of biological oxygen demand over 5 days (BOD_5) to chemical oxygen demand (COD) of thin sludge as a common criterion for sludge stabilization

Ratio of BOD_5 to COD for thin sludge after stabilization	Degree of stabilization	Remarks
< 0.10	Fully stabilized	Only achievable with properly performed digestion of prolonged duration ≫ 20 days
0.10–0.149	Sufficiently stabilized	Normally achievable with properly performed digestion
0.15–0.179	Conditionally stabilized	Normally achievable with properly performed aeration
0.18–0.25	Partly stabilized	–
> 0.25	Non-stabilized	Biological sludge without any degree of stabilization or sludge with aborted stabilization process

There are different features to describe the quality of sludge stabilization depending on the methods of stabilization, which are mentioned below.

However, there is one commonly valid criterion for sludge stabilization, i.e., the ratio of biological oxygen demand over 5 days (BOD_5) to chemical oxygen demand (COD). Depending on the process, different values are stipulated to characterize the degree of stabilization, as shown in Table 3.3 (DWA 2020).

3.2.1 Stabilization by Digestion

Table 3.4 stipulates the specifications of digester operation and characteristics of thin sludge leaving the digester for good digester operation (WEF 2017).

For digested sludge, two parameters are essential for evaluating the degree of stabilization: The reduction or degradation of VDM and the volatile fatty acids content, expressed as acetic acid equivalent. Table 3.5 shows the quality of stabilization depending on these parameters (DWA 2020 and own statement).

The Van Kleeck formula for the calculation of the reduction in VSS is given by:

$$\%\text{VSS reduction} = (\text{VSS}_{\text{in}} - \text{VSS}_{\text{out}}) \times 100/[\text{VSS}_{\text{in}} - (\text{VSS}_{\text{in}} \times \text{VSS}_{\text{out}})] \tag{3.4}$$

where $\text{VSS}_{\text{in/out}}$ is expressed in kg VDM/kg $\text{biosolids}_{\text{in}}$.

If the residence time of sludge in the digester is too short due to hydraulic overload and, for example, the methanogenesis process is not carried out sufficiently, the degradation of hydrocarbons will be incomplete, and more fat and protein will be in the dewatered sludge, causing a delay in the drying process. In addition, a high amount of acetone will be transferred to the exhaust air of the dryer.

Table 3.4 Specifications of digester operation and characteristics of thin sludge leaving the digester are stipulated for good digester operation, amended. (WEF 2017)

SRT (solids retention time) in days	15–20
Typical feed percent solids TS in %	4–7
Feeding rate	As uniform as possible, peaks to be avoided
VDM loading rate in lbs/($ft^3 \times$ day) and in kg/($m^3 \times$ day)	0.12–0.16 1.92–2.57
Temperature in °C	35–39 °C In the case of higher wastewater proportion of slaughterhouses and higher fat load, up to 41 °C
Volatile dry mass (VDM): destruction in % of fed VDM	45–55 Normally, > 50% are necessary for sufficient stabilization
pH	6.8–7.2
Ammonium (NH_4–N) as N in mg/L	800–2000
Alkalinity (ALK) as $CaCO_3$ in mg/L	2500–5000
Methane (CH_4) content of biogas in vol%	60–65
Carbon dioxide content of biogas in vol%	35–40
VA (Volatile acids) in mg/L	50–300
Ratio of VA to ALK in mg VA/mg $CaCO_3$	< 0.3
H_2S content in uncleaned biogas	< 200 ppm
Sufficient mixing of digester while avoiding short-circuit sludge flows reducing SRT	Mixing by external pumps: > 12–14 × volume of digester/day
	Mixing by impeller: specific power > 3.4 W/m^3 of digester volume
	Mixing by gas injection: > 2 m^3 of gas/(m^2 of biggest digester section × h)

Table 3.5 Quality parameters for digested sludge with regard to stabilization

Degree of stabilization	Volatile fatty acid content in digested thin sludge in mg/L, expressed as acetic acid equivalent	Reduction of VDM according to the formula of Van Kleeck in % (own statement for normal digestion without co-substrates)
Non-stabilized	> 1000	< 44
Conditionally stabilized	–	44–47
Partly stabilized	500–1000	47–50
Sufficiently stabilized	300–500	50–55
Fully stabilized	< 300	> 55

In the case of wastewater treatment with boosted digestion—i.e. using a co-substrate or hydrolysis to increase biogas generation—additional analysis with regard to fat and protein content needs to be carried out for the evaluation of the degree of stabilization.

Peaks of industrial wastewater with high contents of protein and/or fat fed to the digester need to be avoided to achieve smooth operation of digestion. In addition, it is of the utmost importance that hydraulic overloading of the digester is prevented in order to keep the solid retention time at design levels.

Table 3.6 shows specific operation data—values of practice—to demonstrate the achieved deterioration of VSS and gas productions depending on digester feed and retention time.

It is obvious that in practice, digestion at different wastewater treatment plants is carried out under different conditions and shows better or worse results.

In any event, a VDM reduction of 38%, as claimed by the US EPA 503 regulation (United States Environmental Agency) for sufficient hygienic status of digested sludge, is not applicable for the evaluation of a sufficient degree of stabilization when sludge needs to be dried after dewatering step.

According to the DWA Work Sheet M 368 (DWA 2014), the anaerobic total sludge age in sewage treatment plants $\geq$ 100,000 PE (> 10 million gallons per day [MGD]) should lie between 16 and 22 days for one-stage digestion and between 12 and 15 days for two-stage digestion. In sewage treatment plants < 50,000 PE (< 5 MGD), the retention time in the digester needs to be between 20 and 28 days for one-stage digestion and between 14 and 19 days for two-stage digestion. Volumetric loading is not a suitable assessment criterion. If liquid co-substrates are added, the organic sludge load should not exceed 0.08–0.12 kg COD/(kg VDM $\times$ d) (DWA 2014).

It is known that digested sludge dried within a short time period after dewatering causes far less transfer of volatile organic compounds (VOCs) into drying air in comparison to sludge that is stored for a longer time, mechanically stressed, and exposed to oxygen before entering the dryer. While odorous gas components such as methane thiol (CH_4S) is degraded during intermediate storage of sludge, other sulfur-containing gas components such as dimethyl sulfide [$(CH_3)_2S$] or dimethyl disulfide [$(CH_3)_2S_2$] are formed in sludge due to microbiological activity and chemical oxidation processes (Novak et al. 2006). Mechanical stress on sludge cells causes the rupture of cell walls and the formation of carbon acids. Oxygen access to the bulk of the sludge inhibits microorganisms responsible for anaerobic degradation of organic compounds. Thus, proteins, fat, and hydrocarbons are no longer transformed to smaller compounds such as methane, carbon dioxide, or water. Broken pieces of hydrocarbons increase the mass of VOC transferred to the gas phase during convective drying of dewatered sludge stored and subjected to mechanical stress and oxygen exposure. Therefore, VOC transfer to drying air might vary by a factor of up

Table 3.6 Values for digester operation at different wastewater treatment plants (WWTP). (According to publications and information from local operators)

Type of digestion	Size (population equivalent) of WWTP and country	Specific volatile feed in kg VDM/ ($m^3 \times$ day)	Co-substrate feeding/proportion of industrial wastewater (IDWW) in %	Digester volume in m^3, retention time in days, temperature in °C	TSS in/out in %	VSS in/out in %	VDM reduction in %	Specific biogas production in $Nm^3/(m^3$ effective digester volume $\times$ day)
Single stage	≈ 250,000 Germany	n.n.	Co-substrate, no ≈ 29 IDWW	3 × 4750 25 37	n.n.	71/53	53.9	≈ 0.70
Double stage	≈ 2,000,000 Germany	1.80	Co-substrate, yes n.n.	10 × 8000 20 35	5.5 2.8	81.7/65.7	57.1	≈ 1.19
Single stage	≈ 200,000 Germany	1.73	Co-substrate, yes 40 Mass% ($DS_{in} = 6\%$, $VDS_{in} = 85\%$) n.n.	2 × 3000 27 37–39	4.0 3.2 (mixture)	69.3 55.9 (mixture)	59.0	≈ 1.15
Single stage	≈ 300,000 Austria	2.13	Co-substrate, yes 18 Mass% n.n.	2 × 4600 25 38	6.5 n.n.	81.9 68.0	53.0	≈ 1.07
Single stage	≈ 300,000 Germany	1.11	Co-substrate, yes ≈ 66 IDWW	2 × 5000 30 41.5	4.2 2.5	79.4 59.3	62.2	≈ 0.64

Table 3.6 (Continued)

Type of digestion	Size (population equivalent) of WWTP and country	Specific volatile feed in kg VDM/($m^3 \times$ day)	Co-substrate feeding/proportion of industrial wastewater (IDWW) in %	Digester volume in m^3, retention time in days, temperature in °C	TSS in/out in %	VSS in/out in %	VDM reduction in %	Specific biogas production in $Nm^3/(m^3$ effective digester volume $\times$ day)
Single stage	$\approx$ 700,000 Germany	2.00	Solid co-substrate, yes 13 Mass% in DM of feeding n.n.	3 × 22.500 40 (52) 37	10.5 4.7	76.0 62.1	48.3	$\approx$ 0.93
Single stage	$\approx$ 650,000 Switzerland	1.17	Co-substrate, no n.n.	4 × 7250 18.5 38	3.1 1.85	70.0 49.0	58.8	$\approx$ 0.66
Single stage	$\approx$ 3,000,000 Columbia	0.90	Co-substrate, no High proportion of IDWW	6 × 8600 20 n.n.	2.7 n.n.	68.0 53.0	46.9	n.n.
Single stage	$\approx$ 40,000 Germany	2.77	50/50 PS/WAS	1700 17 36	8.0 5.0	58.9 45.7	41.3	$\approx$ 0.59

n.n., Not known

Table 3.7 Factors influencing VOC transfer from sludge to drying air

Influence on VOC transfer to drying air		
Factor	High influence on VOC transfer to exhaust air	Lower influence on VOC transfer to exhaust air
Degree of stabilization (thin sludge after stabilization)	Meaning BOD_5/COB > 0.15 → High	Meaning BOD_5/COB < 0.10 → Low
Content of volatile sulfur high: i.e., > 0.8 mass% relative to DM, medium: 0.4–0.8 mass%, low < 0.4 mass%	> 0.8 Mass%	< 0.4 Mass%
Storage time between dewatering and drying (basis retention 3 days)	≫ 3 Days	< 3 Days
Temperature of sludge during storage (regular range 10 °C–25 °C)	> 20–30 °C	< 10 °C
Mechanical treatment of sludge comprising reloading, piston pumping, or other conveying leading to a cell rupture	High mechanical stress on sludge	Direct conveying of dewatered sludge to dryer with low mechanical stress
Intermediate storage of mechanically stressed sludge	Long intermediate storage period, resulting in high ratios of dimethyl sulfide (DMS) to methyl mercaptan (MM) and dimethyl disulfide (DMDS) to MM (methane thiol) in GC-MS analysis, i.e., DMS and DMDS/MM ≫ 1	No intermediate storage period, resulting in low ratios of dimethyl sulfide and dimethyl disulfide to methyl mercaptan, i.e., DMS/MM and DMDS/MM < 1

MM, methyl mercaptan (methane thiol); *DMS*, dimethyl sulfide; *DMDS*, dimethyl disulfide

to 12. During belt drying of sewage sludge, a VOC transfer of 500–7000 mg relative to DM (of throughput) has been measured (HUBER 2021). Extreme cases of up to 14,000 mg C/kg of DM have occurred. As a consequence, it is of the utmost importance for the reduction of the environmental impact from exhaust air of a sludge dryer to shorten the time from digestion to the drying process to the lowest value possible, as well as to reduce mechanical stress on sludge as far as possible.

In particular, piston pumps with tearing and pressing actions that treat sludge at high pressures of up to 100 bar need to be avoided for sludge transportation, and slow moving screw conveyors or low pressure eccentric pumps need to be applied wherever possible if exhaust air from belt dryers needs to be cleaned. Table 3.7 lists the factors influencing VOC transfer to drying air, i.e., whether their contribution is high or low.

During preparation for printing a new and extended report was issued on the quality of anaerobic stabilization and degradation of COB (DWA 2023).

3.2.2 Stabilization by Partial or Full Aeration

3.2.2.1 Statically Thickened Sludge with No or Partial Aeration

Practical experience showed that with cold digestion without aeration, approximately 2–7% (absolute) destruction of volatiles is possible (Hilliges 2016). Partial aeration means that for a period of up to 4–5 days, aeration is carried out in an unheated tank, but it needs to be borne in mind that the degree of stabilization is fair. Nevertheless, a VDR of 75–80% of dewatered sludge can be achieved and sticking of sludge during drying is reduced.

Cold digestion can be calculated with, for example, 50% values of Fair and Moore (Fig. 3.11) curves at 10 °C and for 10 days as average figures for 1 year. There will be production of 50 NL (normalized liter) of biogas with 60 vol% methane per kg VDM feed, resulting in 0.03 Nm^3 CH_4/kg VDM feed and 0.077 kg COD/kg VDM feed as well as in 0.053 kg VDM destroyed/kg VDM feed or 5.3%.

With a starting VSS feed of 83.6%, one obtains VSS destroyed = 0.053 × 83.6% = 4.4% and remaining VSS of 79.2%.

More destruction is possible with prolonged aeration. For further volatile (protein) destruction (aeration), air is injected into the mixing pump line, as shown in Fig. 3.12.

With aeration of 3.3–3.8 kg air per kg of VDM (7.3–8.4 lbs air per lbs of VDM) and an aeration period of approximately 7–9 days, destruction of volatiles of up to 40% is possible (Landia 2017).

Proof of a sufficient degree of stabilization is provided by measuring the alkalinity (acid-binding capacity), determined according to the European standard EN ISO 9963-1

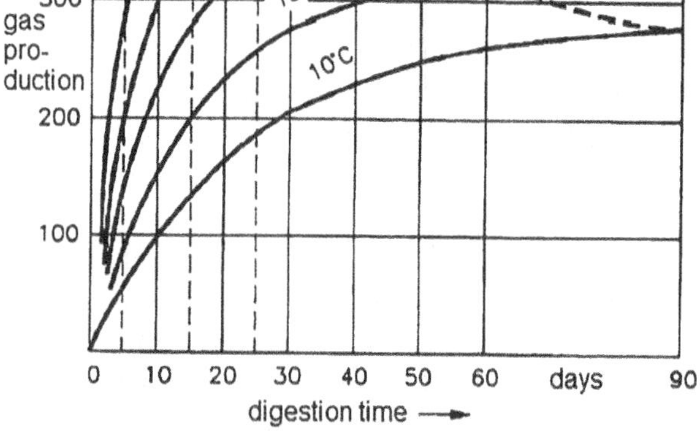

Fig. 3.11 Influence of temperature on digestion of sewage sludge. (Fair and Moore 1937)

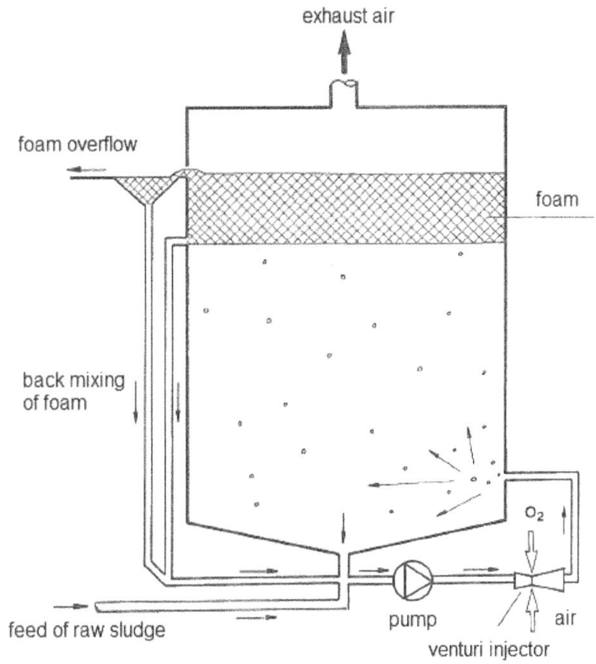

Fig. 3.12 Air injection into sludge tank for further degradation of volatile matter

of February 1996. The values of alkalinity in mmol $CaCO_3/L$ are used as a parameter for the degree of stabilization of the sludge.

- Good stabilization:
  Alkalinity ranges from 15 to 50 mmol $CaCO_3/L$
- Insufficient degree of stabilization:
  Alkalinity of the sludge < 15 mmol $CaCO_3/L$ (sludge cannot buffer enough acids)
  Alkalinity > 50 mmol $CaCO_3/L$ (calcium ions inhibit the acid buffering capacity)

A short description is given in the Appendix for the determination of alkalinity, which is carried out according to the EN-ISO-standard mentioned above.

3.2.2.2 Aerated Sludge (Aerobic Stabilization)

Aerobic stabilization is carried out in smaller wastewater treatment plants with a capacity of less than 20,000 PE in Central Europe. The time and intensity of aeration depends on the organic load of the wastewater and on nitrogen removal. Without nitrogen removal, aeration time ranges from 8 to 20 days. Additional nitrogen removal requires an additional 5 days of oxygen supply.

Sufficient (limited) aerobic stabilization of sludge is verified by testing sludge respiration activity (oxygen depletion) at a temperature of 20 °C (ATV-DVWK 2003). According to the Bavarian State Office for Water Management (*Bayerisches Landesamt für Wasserwirtschaft* 2004), respiration activity is deemed to be the reference value for:

- Sufficient (limited) aerobic stabilization:
  $50\,g\ O_2/(kg\ TSS \times day) < \text{Respiration activity} < 100\,g\ O_2/(kg\ TSS \times day)$
- Aerobically well stabilized sludge shows the following activity value:
  $\text{Respiration activity} < 50\,g\ O_2/(kg\ TSS \times day)$

3.2.3 Effects of a Low Degree of Stabilization on Belt Drying

A low degree of sludge stabilization is always related to:

- High VDR and consequently high protein and fat content in sludge
- Lower DR of dewatered sludge
- Higher content of free water in dewatered sludge and consequently lower stability of extruded sludge strains
- Higher tendency to sticking during mixing of dewatered and fully dried sludge
- Higher need for water evaporation in drying for equal throughput of DM
- Lower specific water evaporation rate due to retarded drying caused by increased protein and fat content and case hardening at the surface of sludge particles
- Increased formation of deposits on heat exchangers/condensers and consequently higher requirements for cleaning and maintenance works

Therefore, with regard to the belt drying process, a high degree of sludge stabilization is always favorable and significantly reduces the expenses/costs of drying.

3.2.4 Practical Analysis of Stabilization of Sludge

Tables 3.8–3.11 provide some analysis of stabilization for different types of sludge (Merten 2021) with corresponding discussions of results.

The samples of digested thin sludge and dewatered sludge were taken after digestion time, i.e., approximately 5 weeks after sampling of sludge before digestion. The stabilization was performed well with regard to volatile acids, as no values exceeded 500 mg/L (see Sect. 3.2.1). The BOD_5/COD values, however, show sufficient to conditional stabilization by anaerobic digestion. The degradation of VDM ranges from 31.1% over 25.4 to 32.2% and confirms that stabilization is not even conditional. This example demonstrates that the stabilization criteria of volatile acetic acid alone is not sufficient for an evaluation of the degree of stabilization of digested sludge.

Table 3.8 Analysis of sludge stabilization for a 12,000 population (P) wastewater treatment plant (WWTP) with digestion. (Merten 2021)

Type of sludge	Thin sludge before digestion	Thin sludge after digestion	Dewatered sludge
TSS/DR in %	2.50–3.89–3.96	2.29–2.21–2.50	21.52–21.53–21.43
VSS/VDR in %	84.30–80.18–83.73	64.30–63.92–63.34	67.23–66.46–65.32
pH	6.09–6.29–6.35	7.33–7.43–7.42	8.71–8.52–8.45
BSB_5/CSB	0.500–0.354–0.209	0.116–0.200–0.144	–
Ammonium in mg/L	96.75–43.35–95.75	113.50–113.00–91.80	In g N/kg DM: 9.04–10.19–7.77
Volatile acids as acetic acid equivalent in mg/L	975–2863–987	366.50–208.00–276.50	–

Digested sludge WWTP 12,000P, values for January to March to May (three samples), no co-substrate in digestion, approximate digestion time of 40 days

The increase in pH is around 1 due to large digesters with a height of approximately 20 m. The BOD_5/COD ratios < 0.200 show sufficiently stabilized digested sludge. Again, the volatile acid values are around 300 mg/L right at the border to fully stabilized sludge. The degradation of volatiles calculated by the Van Kleeck equation is 41.6%–43.1%–39.6%, resulting in only conditional stabilization.

The BOD_5/COD ratio indicates that this sludge is only partly stabilized. The values of thin sludge before aerobic stabilization and after do not match, as samples were taken at one point in time and the values after stabilization belong to a sludge taken 1 month earlier. Whatever the case may be, the fair degree of dewatering of only 15% DR of dewatered sludge additionally demonstrates that stabilization could have been carried out better. Nevertheless, the VDR values are comparatively low in comparison to the degree of stabilization.

Table 3.9 Analysis of sludge stabilization for a 275,000 population equivalent (PE) wastewater treatment plant (WWTP) with digestion. (Merten 2021)

Type of sludge	Thin sludge before digestion	Thin sludge after digestion	Dewatered sludge
TSS/DR in %	4.54–3.76–3.31	2.63–2.22–2.66	27.06–27.78–26.34
VSS/VDR in %	80.38–80.64–78.25	56.75–56.37–56.07	57.74–56.71–56.08
pH	5.89–6.82–6.61	7.44–7.44–7.52	8.54–8.53–8.61
BOD_5/COD	0.209–0.284–0.536	0.144–0.130–0.181	–
Ammonium in mg/L	100.40–75.90–101.00	92.30–91.70–97.90	In g N/kg DM: 5.40–5.53–6.02
Volatile acids as acetic acid equivalent in mg/L	2863–987–1091	312–313–306	–

Digested sludge WWTP 275,000 PE, values for January to March to May (three samples), no co-substrate in digestion, 29% industrial wastewater, approximate digestion time of 35 days

Table 3.10 Analysis of sludge stabilization for a 14,000 population equivalent (PE) wastewater treatment plant (WWTP) with aerobic stabilization of 23–25 days. (Merten 2021)

Type of sludge	Thin sludge before aerobic stabilization	Thin sludge after aerobic stabilization	Dewatered sludge
TSS/DR in %	0.58	2.11	15.12
VSS/VDR in %	67.49	71.41	71.04
pH	6.61	6.37	6.93
BOD_5/COD	0.226	0.243	–
Ammonium in mg/L	9.59	46.45	In g N/kg DM: 5.83
Volatile acids as acetic acid equivalent in mg/L	48.3	150.7	–

Aerobically stabilized sludge WWTP 14,000 PE, values for July, aerobic stabilization for approximately 25 days

Digestion is more complicated in the case of co-substrate application, especially with peaks in co-substrate feedings. To be able to compare values of samples 1–4, one needs to bear in mind that value 1 of thin sludge before digestion corresponds to value 2 after digestion due to the fact that samples of thin sludge before and after digestion were collected at the same time. However, with regard to volatile acids, it can be stated that digestion was carried out sufficiently. Fluctuations in BOD_5/COD of digested thin sludge are large, showing that the stability of the digestion process is not ensured over the whole period of time. The degree of stabilization changes from fully stabilized to non-stabilized with regard to the BOD_5

Table 3.11 Analysis of sludge stabilization for a 200,000 population equivalent (PE) wastewater treatment plant (WWTP) with digestion and a co-substrate for digestion from the food industry. (Merten 2021)

Type of sludge	Thin sludge before digestion	Thin sludge after digestion	Dewatered sludge
TSS/DR in %	3.14–5.24–5.90–4.62	2.23–2.46–2.42–2.71	28.42–27.34–30.22–29.17
VSS/VDR in %	84.25–76.21–79.38–79.62	60.87–60.28–60.20–56.42	63.40–65.57–61.64–62.66
pH	6.15–6.52–6.13–5.97	7.75–7.72–7.93–7.86	8.51–8.39–8.65–8.23
BOD_5/COD	0.215–0.533–0.397–0.555	0.295–0.099–0.424–0.228	–
Ammonium in mg/L	26.1–14.5–43.6–35.8	12.1–35.6–11.9–24.2	In g N/kg DM: 8.82–8.36–10.40–7.08
Volatile acids as acetic acid equivalent in mg/L	10,650–10,883–8038–7642	335–328–331–326	–

Digested sludge WWTP 200,000 PE, values for January to June (four samples), co-substrate in digestion, approximately 40 mass%, approximate digestion time of 21 days (32 °C)

to COD ratio. One reason might be that digester volume is limited and peaks of co-substrate feeding create problems with stabilization. The increase in pH when comparing thin sludge with dewatered sludge is related to CO_2 stripping during the dewatering process. The DR of dewatered sludge is high, resulting in good extrusion behavior for dewatered sludge and sufficient stability of extruded strains during the drying process.

3.3 Specific Properties of Dewatered Sludge

3.3.1 Portion of Industrial Wastewater Fed to a Wastewater Treatment Plant

All properties mentioned are related to municipal sewage sludge (biosolids) from wastewater treatment plants with a portion of industrial wastewater intake of less than 20–25 mass%. Industrial sludge might have considerably different properties due to different constituents and steps of wastewater treatment.

A wastewater proportion—expressed as PE—of more than 30% fed to a municipal wastewater treatment plant results in a sludge that is significantly different to normal municipal sludge. Such sludge more closely resembles industrial sludge.

3.3.2 Density of Dewatered Sludge

A distinction needs to be made between material density and bulk density. A well-stabilized sludge, dewatered by a screw press or a centrifuge, has a crumbly structure and a bulk density of around 700 kg/m^3 (43.7 lb/ft^3). The material density itself depends on water content and proportion of different constituents of sludge such as fat, protein, hydrocarbons, and nonvolatile matter and ranges from 1020 to 1080 kg/m^3 (63.7 to 67.4 lb/ft^3). Higher contents of DM increase material density. As soon as sludge is compressed, for example, in pumps, higher material densities above 1100 kg/m^3 (68.7 lb/ft^3) are possible. With increasing mechanical stress on dewatered sludge, the initial crumbly structure is completely lost and, consequently, bulk equals material density.

3.3.3 Dry Residue and Consistency of Dewatered Sludge

A wide range of DR can be encountered in dewatered sludge depending strongly on the type and degree of stabilization, mode of dewatering, and specific amount of polymer used for sludge flocculation. The lowest DR values are around 14–15% for non-stabilized

Table 3.12 Sludge consistency depending on DR value

DR in %	Consistency
< 15	Liquid, pumpable with fluid pumps
15–25	Sticky phase in solar drying
25–35	Still malleable, slurry, smearing, set, pumpable with eccentric-screw and piston pumps
40–60	Sticky phase in thermal drying and in general when high mechanical force applied, pasty, early granulation > 45%, pumpable in piston pumps up to a DR of maximum 45%
60–65	Granular, spreadable
> 80–85	Granular, dusty

sludge dewatered with belt presses. The highest DR values are around 30–34% in the case of properly digested sludge dewatered with centrifuges or screw presses. Thermally or thermochemically hydrolyzed sludge can be dewatered to up to 40% DR as the sludge cells are ruptured by steam or thermochemical treatment. Hydraulic filter presses are the dewatering machines most commonly used. The dewatered sludge is often referred to as 'sludge cake' in English-speaking regions.

The DR value determines sludge consistency, as described in Table 3.12.

The consistency of dewatered sludge changes dramatically when mechanical stress is applied by frequent loading and unloading actions or long distance pumping/mechanical conveyance. The granular structure is completely destroyed and a pasty, sticky mass that sometimes behaves much like custard can be encountered.

3.3.4 Specific Heat Capacity of Sludge

The specific heat capacity of the DM of sewage sludge is around 1.26 kJ/(kg DM × K). Assuming the specific heat capacity of water to be on average 4.19 kJ/(kg × K)—ranging from 4.228 for 0 °C to 4.217 kJ/(kg × K) for 99.6 °C—heating and cooling processes applied to dewatered sludge can be estimated with the DR in % by using:

$$c_{\mathrm{p}} = \left(1 - \frac{\mathrm{DR}}{100}\right) \times 4.19\,\mathrm{kJ/(kg \times K)} + \left(\frac{\mathrm{DR}}{100}\right) \times 1.26\,\mathrm{kJ/(kg \times K)} \tag{3.5}$$

3.3.5 Heat Conductivity of Sludge

Heat conductivity λ_{C} of sludge in W/(m × K) according to Kranert (1988):

$$\lambda_{\mathrm{C}} = \frac{1}{1.721966303 + 0.002386597 \times (\mathrm{DR} \times 100)^2} \tag{3.6}$$

with DR in %.

Table 3.13 Lower heating value (LHV) of volatile dry mass (VDM) of different types of sludge

Sludge type	LHV range in MJ/kg VDM
Primary sludge (PS)	23.2–32.5, typically 25.4
Waste activated Sludge (WAS)	20.9–23.2, typically 21.8
Mixed sludge (PS + WAS)	Typically 23.4
Digested sludge	18.6–25.5, typically 24.6
Screenings	20.0–24.4

3.3.6 Lower Heating Value (calorific value)

The lower heating value (LHV) of dewatered sludge or biosolids depends on VDR and consequently on the type of sludge and its degree of stabilization. For example, primary sludge can be expressed by the formula $C_{23}H_{35}O_8N$ (carbon content 60% in DM) and biological sludge (WAS) by $C_{13}H_{24}O_6N$ (carbon content 53% in DM; Barber 2020). Also, intermediate storage of sludge results in degradation of VDM and in a decrease in LHV. Typical LHVs range from 17,000 to 23,000 kJ/kg VDM. Sludge constituents, for example, hydrocarbons, have a calorific value of around 17,000 kJ/kg, proteins of around 23,500 kJ/kg, and fats of around 39,500 kJ/kg (Leschber and Loll 1996). Tests carried out by HUBER SE on digested dewatered sludge showed calorific values/LHV of 19,500 kJ/kg VDM at a German WWTP of 300,000 P, and 21,500 kJ/kg VDM at an Austrian WWTP of 400,000 P with co-digestion of food waste (HUBER SE 2018b). Reimers and Englande indicate an LHV depending on the type of sludge, as shown in Table 3.13 (Reimers and Englande 1983; Barber 2020).

3.3.7 Properties and Constituents of Municipal Dewatered Sludge and Municipal Thin Sludge

All the constituents of sludge indicated in Table 3.14 are expressed in % relative to DM or in mg/kg of DM unless not otherwise stated. The data are explained in the 'Remarks' column, and in some cases the effect of the constituents on subsequent process stages such as dewatering, drying, or incineration are described. Some values are related to municipal thin sludge or wastewater.

Table 3.14 Sludge constituents and characteristics (Bank 1995; Bayerisches Landesamt für Wasserwirtschaft 2000; DWA 2015, 2019; Loll 1992; Roediger 2014; Thomé-Kozmiensky 1998; Wiesler 2009; Wikipedia 2022, 2022)

Sludge constituent/ characteristic	Value range	Remarks
Loss on ignition (organic/volatile matter) VDR	Digested (anaerobically stabilized) sludge: average VDR 58%, range from 50 to 62%; aerobically stabilized mixed sludge: average VDR 68%, range from 60 to 72%; waste activated sludge: range of VDR from 80 to 87% Mixed sludge primary/waste activated sludge: range of VDR from 72 to 80%	WAS and mixed PS/WAS with high protein content, low achievable degree of dewatering and high susceptibility to sticking and plate formation during drying
Total sulfur	0.8–2.0 (3.1)% Relative to DM	Low values for normally digested sludge, average values of 1.2–1.8% for digested sludge and different addition of co-substrates to digestion, higher value for aerobically stabilized sludge. Extreme values of more than 3% result from wastewater of anaerobic status in low-flow sewage
Sulfate (SO_4^{2-})	Municipal wastewater: 50–80 mg/L (Koppe and Stotzek 1999) Industrial wastewater: textile finishing: 750–1050 mg/L Oil and fat finishing: Up to 12,000 mg/L Yeast fabrication: up to 3200 mg/L Metal surface treatment with sulfuric acid: up to 7200 mg/L Refineries for dietary fats: up to 1000 mg/l (Weismann and Lohse 2007) Sulfate values of dewatered sludge range from 150 up to 4000 mg/kg DM	High values in the case of a high proportion of special industrial wastewater; values above 500 mg SO_4^{2-}/L result in a decrease in biogas production in digestion due to high H_2S development (Meinck et al. 1968); long-term values of sulfate > 400 mg SO_4^{2-}/L in wastewater lead to corrosive attack on cement constructions (ATV 1985) Higher values of sulfate in dewatered sludge are associated with digested sludge
Volatile sulfur (organic sulfur like methyl mercaptan or dimethyl sulfide)	0.1–1.5% Relative to DM, average values of around 0.6–0.7%	Roughly calculated by reducing total sulfur value with sulfate sulfur value. Influences odor units in exhaust air of dryers
Calcium (Ca)	3–10% Relative to DM	Ca is oxidized to CaO during incineration, which can lead to the formation of eutectic compounds with other elements and a decrease in the ash melting temperature. The formation of ash agglomerations disturbs the fluidized bed

Table 3.14 (Continued)

Sludge constituent/ characteristic	Value range	Remarks
Magnesia (Mg)	0.6–1.7% Relative to DM	Increases fertilizer value of sludge ash
Potassium (K)	0.3–0.8% Relative to DM	K is oxidized to K_2O during incineration, which can lead to the formation of eutectic compounds with other elements and a decrease in the ash melting temperature. The formation of ash agglomerations disturbs the fluidized bed
Iron (Fe)	0.5–12.0% Relative to DM	Iron content in dewatered sludge depends on iron feed for phosphate precipitation with, for example, $FeClSO_4$ and sulfur binding in digestion with the addition of $FeCl_2$ or $FeCl_3$. Average values with Fe dosing for phosphate precipitation of 5–7%, without phosphate precipitation, 1.5–2.0% (EAWAG 2013). Increases susceptibility to self-heating in bulks of dried sludge during storage in closed silos
Extractable ammonium (NH_4^+–N)	0.2–2.0% N relative to DM	Dependent on protein content of wastewater or co-substrate feed to digestion. High values result in high emission of NH_3 gas during drying at pH-values of dewatered sludge exceeding 8
Total nitrogen (N)	Typically 2–6%/average values around 5%, maximum values around 9% relative to DM	For digested sludge usually around 4–5%
Protein value = Kjedahl nitrogen content × 6.25 relative to DR	18–53%, Typically 25–35%	High values occur with insufficiently digested or aerated sludge; very low values for aerobically stabilized sludge and wastewater treatment plants without or with very low industrial wastewater feed. With digested sludge, protein values range from 26 to 29%; sticking behavior during drying at VDR of 70% starting with protein value of 42%, at 75% VDR starting with 37%, at 80% VDR starting with 34%

Table 3.14 (Continued)

Sludge constituent/ characteristic	Value range	Remarks
Phosphorus (P)	Typically 0.3–6.0% relative to DM In Germany: WWTP of up to 50,000 PE: 2–2.5%, over 50,000 PE: 2.5–3.5%	Limit for economic recycling of P from sludge ash for spray granulation process: around 14–15% P_2O_5 or 6.1–6.5% P in ash (corresponding to approximately 2.1–2.3% P in DM of sludge assuming 65% VDR in dewatered sludge) 2014: Average value of P in German sludge ashes ≈ 7.3%, alternating from 1.5 to 13.1% (Krüger and Adam 2014)
Vivianite [$Fe_3(PO_4)_2 \times 8H_2O$]	2–20% Relative to DM	Higher values in the case of digested sludge, lower in the case of aerated sludge. Vivianite accelerates at temperatures of dried sludge > 80 °C in silos; susceptibility to self-ignition in beds of dried sludge
Chloride (Cl) (UBA 2013; GWI 2012)	0.05–0.8% Relative to DM, average values around 0.2–0.5% for municipal sludge	High chlorine contents results in formation of higher dioxin and furan contents in exhaust gas of sludge incineration in the case of lower sulfur contents of sludge. Chlorine supports formation of sulfur trioxide, thus increasing the dew point temperature of exhaust gases and danger of corrosion in exhaust gas cleaning systems (DWA 2011)
Fluor (F) (UBA 2013; GWI 2012)	0.008–0.035% Relative to DM	Promotes corrosion on surfaces and in fissures of incineration plants
Mercury (Hg)	0.4–2.6 mg/kg DM	Dependent on proportion of special industrial wastewater fed to wastewater treatment plant
Cadmium (Cd)	0.7–1.3 mg/kg DM, maximum 4.5 mg/kg DM	Dependent on proportion of special industrial wastewater fed to wastewater treatment plant
Arsenic (As) (Koppe and Stotzek 1999; UBA 2017)	4–30 mg/kg DM, typically ≈ 10–20 mg/kg DM	Dependent on proportion of special industrial wastewater fed to wastewater treatment plant
Sand content (insoluble in HCl acid), Germany (DWA 2008)	85% Percentile*, relative to VDR Digested sludge: 47.5% (range 2.1–74%), relative to VDR of 58%: 27.7% Aerobically stabilized secondary sludge: 53.2% (range 1.5–72%), relative to VDR of 68%: 36.2%	* 15% of values exceed this value. Causes mechanical wear in sludge conveying or mixing systems and extruders

Table 3.14 (Continued)

Sludge constituent/ characteristic	Value range	Remarks
pH-Value of dewatered sludge	Usual values for digested sludge: 7.2–7.8, for digested sludge from digesters higher than 15–20 m, CO_2 stripping during dewatering needs to be considered: increase in pH-value up to 8.2–8.5; for non-stabilized or minor stabilized sludge of high content of volatile acids: 6.0–7.0	Dependent on proportion of wastewater of industrial origin and on sulfur content of wastewater. High pH-values of dewatered sludge increase ammonia transfer into dryer exhaust air, while low values promote transfer of odorous sulfur components
Chemical oxygen demand (COD) total dissolved in thin sludge (approximate values) (Schaum et al. 2016)	Primary sludge: $\approx$ 50,000 mg/L Waste activated sludge: $\approx$ 30,000 mg/L Digested sludge: $\approx$ 25,000 mg/L Flotation sludge (industrial): $\approx$ 160,000 mg/L	
Volatile organic acids in thin sludge as acetic acid equivalent (Fresenius et al. 1984)	Fresh sludge from mechanical wastewater treatment (primary sludge [PS]): 1800–3600 mg/L Sludge from trickling filters (WAS): 1800–3600 mg/L Improperly digested sludge (acid fermentation): 2500–4000 mg/L Fairly digested sludge: 1000–2500 mg/L Properly digested sludge: 100–300 mg/L	High contents of volatile organic acids indicate a low degree of sludge stabilization. Usually even properly stabilized dewatered sludge stored for a longer period of time shows higher contents and lower pH-values than the original sludge. Proper digestion of municipal sludge results in a low volatile organic acids content of less than 500 mg/L or in a TS content of thin sludge of 2.5% of less than 0.02 mg/kg DM
Elementary analysis of sludge (approximate ranges, values), valid for digested sludge, WAS, and mixed sludge	Carbon (C): 50–54% Hydrogen (H): 6–8% Nitrogen (N): 5–7% Sulfur (S): 0.8–3.1% Oxygen (O): 31–36% Exception: primary sludge (PS) with nitrogen (N) content of 3–5%	Relative to VDM (volatile dry mass; free of water and ash)

References

Abu-Orf Mm Goss T (2012) Comparing Thermal Hydroysis Processes for Solids Pretreatment Prior to Anaerobic Digestion. Residual and Biosolids. https://www.researchgate.net/profile/Mohammed-Abu-Orf/publication/272221844_Comparing_thermal_hydrolysis_processes_CAMBI_and_EXELYS_for_solids_pretreatment_prior_to_anaerobic_digestion/links/54f09c200cf2f9e34efce360/Comparing-thermal-hydrolysis-processes-CAMBI-and-EXELYS-for-solids-pretreatment-prior-to-anaerobic-digestion.pdf

ATV (1985) Wastewater Technology VII Industrial Wastewater with Inorganic Ingredients (In German), 3rd edn. Ernst & Sohn, Berlin

ATV-DVWK (2003) Deutsche Vereinigung für Wasserwirtschaft, Abwasser und Abfall e. V. (German Water, Wastewater and Waste Association), Regional Association Bavaria. Guidelines No. 2–3, Operation of wastewater plants: Proof of aerobic sludge stabilisation via respiration activity (In German). Munich

BAFU (2009) Schweizerische Eidgenossenschaft Bundesamt für Umwelt Recovery of Phosphorous from wastewater cleaning – a state of the art (In German) 29/09. BAFU, Bern

Bank M (1995) Basic knowledge of environmental protection technique, 3rd edn. Vogel, Würzburg (In German)

Barber B (2020) Webinar CAMBI thermal hydrolysis for optimizing biosolids incineration

Baudirektion Kanton Zürich (2013) State of the Art of Mechanical Dewatering of Sewage Sludge (in German). Amt für Abfall, Wasser, Energie und Luft (www.awel.zh.ch)

Bavarian State Office for Water Management (2004) Work Sheet No. 4.7/11 Proof of stabilisation criteria with aerobic sludge stabilisation

Bayerisches Landesamt für Wasserwirtschaft (2000) Heavy metals in municipal waste water and sewage sludge (In German) Merkblatt 4.6/1 v. 03.11.2000

CAMBI (2022) Email correspondence with CAMBI enterprise, Asker, Norway. https://www.cambi.com/what-we-do/thermal-hydrolysis/

CNP CYCLES (2021) Details on Grain Size Distribution of stuvite crystalls and on two stage AirPrex® Macro Process. Karlstein am Main. https://cnp-cycles.de

DWA (2008) Merkblatt DWA-M 383 Characteristic Values of Sludge Dewatering (In German). DWA, Hennef

DWA (2011) DWA-M 386 Merkblatt: Thermal Treatment of Sludge – Mono Incineration (In German). DWA, Hennef

DWA (2014) Work Sheet 368 Biological stabilisation of sewage sludge (In German). DWA, Hennef

DWA (2015) Theme Issue: Anaerobic or common aerobic stabilisation (In German). DWA, Hennef

DWA (2019) Merkblatt M 383 – Specific figures of sludge dewatering (In German). DWA, Hennef

DWA (2020) Parameters for evaluation of product quality of biologically stabilized sludge. Working paper of DWA-working group KEK 2.1 (in German). KA Korrespondenz Abwasser Abfall 67(12):969

DWA (2023) Parameters for the quality of Anaerobic Stabilization and COD degration. Work Report of DWA Working Group KEK-2.1. KA Korrespondenz Abwasser Abfall 70(7):525–531

EAWAG (2013) Phosphate Depreciation with Alumina Products – Clarification of Feasibility (In German). Kantonale Tagung für das züricherische Klärwerkspersonal Opfikon, 18.–20.November 2013

Fair GM, Moore EW (1937) Observation on the Digestion of a Sewage Sludge over a Wide Range of Temperatures. Sewage Work J 9(1):3–5

Fresenius et al (1984) Wastewater technology. Springer, Berlin

GWI Sludge Management 2012

Heindl A (2016) Photo of a CAMBI Sludge processing unit at Crossness Wastewater Treatment

Heinzmann B, Lengemann A (2014) From a Problem of Operation to the Product "Berliner Pflanze". P-Recovery in the Wastewater Treatment Plant Waßmannsdorf (In German). Workshop 11th of June 2014: Phosphorous in Agriculture – Strategies for an Ending Resource. Leibniz Institute for Agricultural Technique, Potsdam-Bornim

Higgins MJ (2016) Effect of Soluble P on Bioflocs and Dewatering. Sludge Tech Conference, 26.–28.06.2016

Hilliges R (2016) Demands on Sludge Stabilization on waste water treatment plant (In German). Bayer. Landesamt für Umwelt, pp 76–167

HUBER SE (2016) Photo of digested sludge

HUBER SE (2017) Photo of aerobically stabilized sludge

HUBER SE (2018a) Photo of mixed sludge (PS + WAS)

HUBER SE (2018b) Ordered Measurements of Heating Values of Sludge Samples

HUBER SE (2019) Photo of a HUBER belt dryer at its discharging end

HUBER SE (2021) Measurements of HUBER SE during practical operation of belt drying of sewage sludge

HUBER US (2018) Exhaust air measurements on a belt dryer treating mixed sludge MS

KEMIRA (2020) Kemira_water_NL_2016-3-sludge-dewatering-chemicals_Monsanto_case.pdf

Kopp J (2017) Optimization of Dewatering of Sewage Sludge (In German). 49. Aargauische Klärwärtertagung, Suhr, 9 Nov 2017

Koppe P, Stotzek A (1999) Municipal Wastewater (In German), 4th edn. Vulkan, Essen

Kranert M (1988) Releasing and Using of thermal Energy During Sludge Composting (In German) Freisetzung und Nutzung von thermischer Energie bei der Schlammkompostierung. Stuttgarter Berichte zur Abfallwirtschaft, vol 33. Erich Schmidt Verlag, Bielefeld

Kroiss H, Svardal K (2009) Energy demand of wastewater cleaning plants (In German). ÖWAW 11–12:170–172

Krüger O, Adam C (2014) Ash of Sewage Sludge: Potentials for Recovery of Reusable Materials (In German). VDI Wissensforum für Klärschlammbehandlung, Straubing, 18 Sept 2014 Bundesanstalt für Materialforschung, In

Kunz P (1992) Umwelt-Bioverfahrenstechnik (Environmental bio-process-engineering). Vieweg, Braunschweig, Wiesbaden

Landia (2017) Information on Application of Jet Pumps for Aeration. Landia a/s DK-6940 Lemst, Denmark

Leschber R, Loll U (1996) ATV-Handbuch Klärschlamm. Ernst & Sohn, Berlin. 4. Auflage

Loll U (1992) Recycling of Sludge 3 – Processing and Treatment of Sludge (In German). EF-Verlag, Berlin

Meinck F, Stoof H, Kohlschütter H (1968) Industrial Wastewater (In German), 4th edn. Gustav Fischer, Stuttgart

Meinel F, Gnirss R (2014) Elimination of Phosphorous: Ecological Balance and Comparison of Economic Efficiency (In German). WWT 10:28–31

Merten M (2021) Influence of Sludge Stabilization on Extrusion and Drying of Dewatered Sludge (In German). Master Thesis Technische Hochschule Nürnberg Georg Simon Ohm

Novak JT et al (2006) Generation Pattern of Sulfur Containing Gases. Water Environ Res 78(8):821–827

NRW (2018) Energy in wastewater Treatment Plants (In German). Manual of Nordrhein-Westfalen NRW, 2nd edn. (www.umwelt.nrw.de)

Ohlinger KN, Young TM, Schroeder ED (1998) Predicting Struvite Formation in Digestion. Water Res 32(12):3607–3614

Panico A et al (2014) The Effect of Substrate-Bulk Interaction on Hydrolysis Modeling in Anaerobic Digestion Process. Sustainability 6:8348–8363

Peeters B (2011) Effect on dewatered sludge composition on its dewaterability and sticky phase. Dissertation Catholic University of Leuven, Faculty of Engineering.

Ramadhania N (2021) Investigation on Phosphorus Recovery from Wastewater Sludge as Struvite at SNJ Wastewater Treatment Plant (IVAR): A Pilot-Scale Study. Master Thesis. University of Stavanger, Norway

Reimers RS, Englande AJ (1983) Chemical Properties. In: Sludge Characteristics and Behavior. NATO ASI Series E: Applied Sciences No.66. Martinus Nijhoff Publishers, Boston The Hague Dordrecht Lancaster

Roediger M (2014) Sludge Digestion or common Aerobic Stabilization on Wastewater Treatment Plants of Small or Medium Size (In German). 26. Hamburger Kolloquium zur Abwasserwirtschaft, 23 Sept 2014

Schaum C et al (2016) COD-Analysis of Sludge – Results of Comparing Measurements (In German). Ka Korrespondenz Abwasser Abfall 63(4):291–298

Thomé-Kozmiensky KJ (1998) Disposal of Sewage Sludge (In German). TK-Verlag, Neuruppin

UBA (2013) Umweltbundesamt – Disposal of Sewage Sludge (In German). Umweltbundesamt, Dessau

UBA (2017) Umweltbundesamt – Arsene in Wastes (In German). Texte 113/2017. Umweltbundesamt, Dessau/Roßlau. https://www.umweltbundesamt.de/sites/default/files/medien/1410/publikationen/2017-12-14_texte_113-2017_arsen-in-abfaellen.pdf. Accessed 04-08-2022

WEF (1995) Manual of Practice No. 22 Odor Control in Wastewater Treatment Plants. Water Environment Federation, Alexandria

WEF (2017) US Water Environment Federations fact sheet "Anaerobic digestion fundamentals". https://www.wef.org/globalassets/assets-wef/direct-download-library/public/03---resources/wsec-2017-fs-002-mrrdc-anaerobic-digestion-fundamentals-fact-sheet.pdf. Accessed 25 Sept 2020

Weismann D, Lohse M (2007) Sulfide-Practical Manual of Wastewater Technique (In German). Vulkan, Essen

Wiesler F (2009) Benefits and Risks of Fertilization with Sludge (In German). Presentation LUFA Speyer 53. Kreuznaher Wintertagung.

Wikipedia (2022) https://de.wikipedia.org/wiki/Calciumacetat. Accessed 24 Aug 2022

Technological and Economic Importance of Sludge Dewatering 4

With regard to the drying of dewatered sludge, optimization of sludge dewatering helps tremendously in reducing energy demand, especially at lower dry residue (DR) contents. Mechanical water separation requires only around 0.2% of the specific energy demand for thermal drying in terms of kWh/kg of water removed. Increasing DR by mechanical dewatering results in a decrease in water evaporation for the subsequent drying process, which is shown in Fig. 4.1 for a dry mass (DM) throughput of 500 kg/h and a final DR content of dried sludge of 90%. The water evaporation for a sludge of 25% DR (1944.4 kg/h) is 500 kg/h lower than that of a 20% DR sludge (1444.4 kg/h). The reduction in water evaporation starting with 14% and ending with 40% DR is taken into account. Thus, the thermal energy demand—disregarding the higher electric energy demand of the dewater-

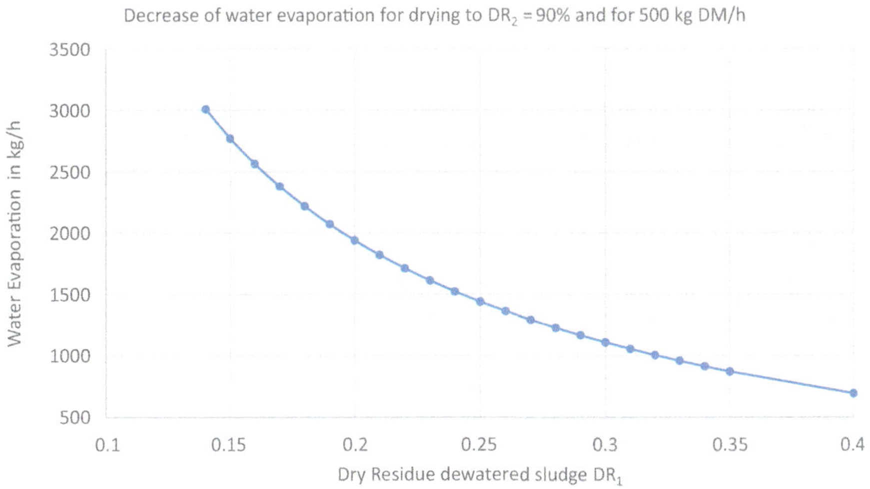

Fig. 4.1 Effect of the dry residue (*DR*) of dewatered sludge on water evaporation of a belt dryer for a constant dry mass throughput of 500 kg/h

A. Heindl, *Belt Drying of Sludge*, https://doi.org/10.1007/978-3-662-66512-1_4

ing unit—will be reduced analogously to the necessary drying area for sludge with the same drying behavior.

Thus, improving the dewatering system before installing a belt dryer will result in considerable savings in CAPEX and OPEX for the drying system, normally well in excess of the additional costs for the reconstruction or modernization of the dewatering system.

Good dewatering requires the aggregation of bacteria into sludge flocs, as well as high values in floc size, floc density, and floc strength (Higgins 2016).

Factors affecting the dewatering of sewage sludge include:

- Type of sludge: primary sludge, mixed sludge (e.g., 1:1 primary sludge to biological surplus sludge; increasing proportion of biological sludge decreases dewaterability), aerated sludge, digested sludge.
- Type of dewatering machine: centrifuges, belt filter presses, chamber filter presses, screw presses (Kopp 2014). Shear forces are different and allow the stability of sludge flocs to be adapted to the dewatering system.
- Ratio of monovalent cations M (Na^+, K^+, NH_4^+) to divalent cations D (Ca^{2+}, Mg^{2+}), i.e., M/D factor < 2 for good dewatering (Higgins 2016).
- Content of extracellular polymeric substances (EPS; proteins, polysaccharides), i.e., high EPS contents decrease the achievable DR of dewatering, 1 g of EPS binds up to 4–5 g of water (Kopp and Ewert 2006)
- The use of biological phosphorus removal in wastewater treatment, i.e., change in digester chemistry and poor floc formulation. As a result, dewatering efficiency is reduced by an absolute DR of 4–5% in relation to normal wastewater treatment with phosphorus removal via, for example, dosing of iron or aluminum, caused by a high phosphate content in thin sludge, thereby enhancing the water binding capacity of EPS. Fig. 4.2 demonstrates the effect of phosphate content on the achievable DR content of dewatered digested sludge of municipal origin (Kopp 2019).
- Fat deriving from co-substrate feeding to digestion for higher biogas generation: usually a reduction of up to minus 3% DR absolute in dewatering (Kopp 2014).
- Sludge temperature, i.e., preheating the sludge up to 35–60 °C might increase the DR of dewatered sludge by up to 25% (relative) and reduce polymer costs by up to 40% (relative; López-Paulau 2015). However, a minimum dosage of polymer is still required and the formation of $CaCO_3$ needs to be prevented, either by lowering the pH-value of thin sludge or by adding an anti-scaling agent to thin sludge as soon as the sludge temperature exceeds 35 °C.
- A flocculating polymer adapted to the nature of the sludge with regard to electric load as well as fiber and protein content.

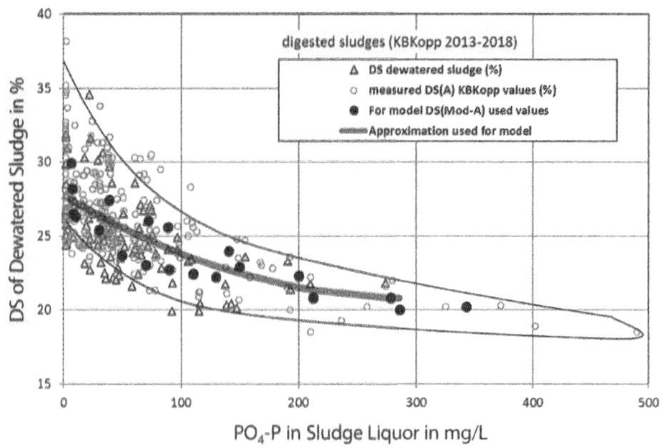

Fig. 4.2 Effect of soluble phosphate content (0.45 μm) of thin sludge on the achievable dry residue of digested dewatered sludge of municipal origin: measurements from 2013 to 2018. (Kopp 2019). *DS* (dewatered sludge) is equivalent to dry residue (DR), as used in this book

- Sufficient maturing time of the flocculant: This is essential to achieve full exploitation of polymer capacity and to produce stable sludge flocs.
- Sufficient mixing energy: This must be used when adding the flocculant to the sludge before dewatering, either by an inline motor-mixing device or using a mixing flap.

The higher the colloidal nature of sludge, the lower the DR achieved in dewatering. Thus, the different types of sludge can be arranged in a line of rising colloidal nature: Primary Sludge → Digested Primary Sludge → Fresh Mixed Sludge → Digested Mixed Sludge → Biological Sludge (SNF Floeger 2003).

The higher the shear forces during the dewatering process, the higher the stability of formed flocs needs to be to obtain sufficient dewatering results. The different types of dewatering equipment can also be arranged according to increasing shear forces: Belt Press → Screw Press → Centrifuge.

When sludge is dewatered, several factors such as type of dewatering equipment, cost of flocculants, and disposal costs influence the degree of dewatering. If the sludge needs to be dried after dewatering, capital and operational costs will also play a decisive role. In countries where disposal costs for dewatered sludge and energy costs for gas and electricity are low, dewatering is not carried out to the highest level as the cost of flocculants prevails. In countries where high disposal costs are common, energy costs are high, and the sludge needs to be dried, dewatering is optimized to save money regarding the entire sludge treatment process.

In any event, a high value of electrical conductivity of municipal sludge always results in higher polymer demand. There are three ranges of electrical conductivity of thin sludge according to Table 4.1.

Table 4.2 provides a number of rules for polymer selection. The vast majority of flocculants used are cationically (+) charged, being able to catch anionically (−) charged organic particles. There are some rules of thumb for the selection of polymers for floc formation, as described in Table 4.2.

Table 4.1 Ranges of electrical conductivity and grading of thin sludge from municipal wastewater treatment plants

Regions of electrical conductivity in mS/cm	Grading
< 1.5	Low
1.5–6	Medium
> 6	High

Table 4.2 Rules of thumb for polymer selection depending on sludge quality

Quality characteristic of the sludge or dewatering system	Polymer selection
Low electrical conductivity	Low electric charge
High electrical conductivity	High electric charge
High organic dry matter	Cross-linked structure
High organic dry matter and low fiber content	Cross-linked structure with low electric charge
High fiber content	Linear, long-chain
Centrifuge dewatering	High to very high molecular weight

Fig. 4.3 describes how polymers behave in the case of high VDR in dewatered sludge in combination with fibers. On the left side, strong bridge formation can be observed for the

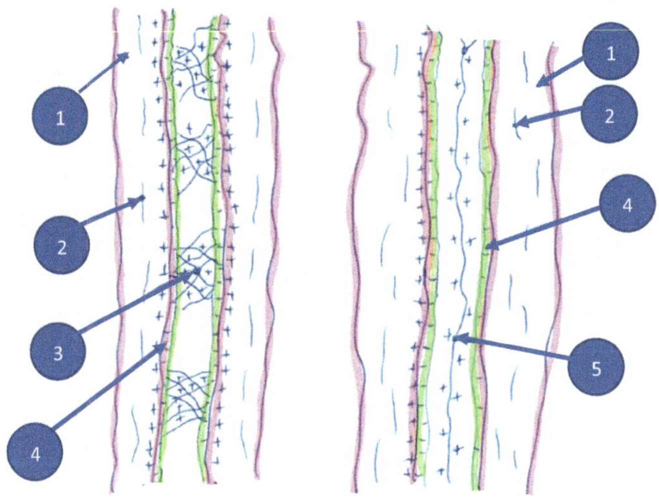

Fig. 4.3 Behavior of polymers for floc formation in the case of dewatered sludge with high organic and protein contents (*left*; strong bridge formation with a strong cationic and cross-linked polymer, *right*: linear and weak polymer with less sticking). *1* Long-chain protein strain with positive surface loading at pH-values < 7 (acid milieu); *2* fibers reinforcing protein strains; *3* cross-linked polymer with high electric charge; *4* liquid surface phase with negative electric charge; *5* linear, long-chain polymer with weak cationic electric charge

use of strong cationic and cross-linked polymers, causing sticking of long-chain protein strains. On the right side, there will be less sticking in the case of a linear and weakly cationic polymer.

Industrial thin sludge might behave completely differently; even with an electrical conductivity exceeding 10 mS/cm, a high degree of dewatering at a medium dosage of polymer can be achieved (DWA 2008).

On the other hand, overdosing of water-soluble organic polymers, i.e., flocculants, not only worsens the economic aspect of sludge dewatering and promotes sticking of sludge, it also effects the efficiency of convective drying. Polymers with high water-binding capacity hinder water diffusion out of sludge particles, thereby slowing down the drying process. Overdosing occurs when the zeta potential measured in mV becomes positive and increases sharply (Kopp 2002). The zeta potential is based on electrophoresis and describes the repellent force of the surfaces of the particles.

A qualitative evaluation of polymer overdosing can be carried out on site (HUBER 2018):

1st Step evaluation: Filtrate liquid (can be free of particles), aspirated by a 0.45-μm paper filter (similar to the determination of filterable matter). If there is virtually no flow through the paper filter, there is a high possibility of polymer overdosing.

2nd Step evaluation: Mix untreated thin sludge with the same volume of filtrate, shake, and stir. If stable flocs are formed, polymer overdosing is indicated.

A semi-quantitative evaluation can be carried out in a laboratory with accurate determination of total solids (TS) and total suspended solids (TSS) of thin sludge (HUBER 2018).

Step 1: Determination of TSS and TS (including dissolved salts) in thin sludge before inline mixing of polymer, calculation of dissolved matter by DR minus TSS.

Step 2: Determination of TS of liquid polymer–water mixture.

Step 3: Determination of TS in solid-free filtrate.

Step 4: Deduction of portion of dissolved matter in filtrate.

Step 5: Mass difference of TS in solid-free filtrate and portion of dissolved matter = solids caused by polymer.

Step 6: Comparison of DR in polymer–water mixture with solids caused by polymer = polymer concentration in filtrate = TS solid-free filtrate—(TS–TSS) thin sludge.

It must be emphasized that choosing the right polymer (low or high molecular weight, linear or branched, cationic or anionic) and the correct method of its conditioning before it is added to sludge is the key to efficient and economic dewatering, and tests with different polymers are essential.

A first qualitative indication for polymer efficiency is the so-called *cup-test*, when polymer is added to sludge and swiveled. The size and stability of flocs resulting from the cup test provide an initial assessment on which flocculant appears to be effective and which not for a given sludge.

As a rule of thumb, approximately 5–15 g of active polymer substance per kg of DM of sludge is used for good dewatering results. Further tests for polymeric flocculants are described by Kim (1995).

It makes sense to optimize polymer application for dewatering to achieve minimum costs by introducing a so-called sludge factor S on the basis of test results integrated in a tender for polymer purchase (Stube and Kopp 2013):

$$S = 100 \times \frac{f_{\mathrm{P}} \times [(M_{\mathrm{p}} \times P_{\mathrm{p}}) + (M_{\mathrm{e}} \times P_{\mathrm{e}})]}{f_{\mathrm{S}} \times [\%\mathrm{DR} \times \mathrm{DM}]} \tag{4.1}$$

with

f_{P}: Factor for the evaluation of costs for polymer consumption per kg of active ingredient, usually 1–2; depending on polymer costs, higher values will result in reducing polymer costs and decreasing DR of dewatered sludge. In 2021, a rough price indication was ≈ € 5.50 per kg of active ingredient for emulsion polymers. Prices strongly depend on polymer variety, annual consumption, and local market conditions.

f_{S}: Factor for the evaluation of costs for sludge disposal, usually 1–2; depending on disposal costs, higher values will result in higher DR contents of dewatered sludge. The higher factor will be applied in the case of planned drying of dewatered sludge.

M_{p}: Quantity of polymer powder used in kg/a for thickening and dewatering.

M_{e}: Quantity of emulsion polymer used in kg/a for thickening and dewatering.

P_{p}: Price of powder polymer in €/kg of active substance.

P_{e}: Price of polymer emulsion in €/kg of active substance.

DM: Dry mass of sludge in metric ton/a.

%DR: Dry residue of dewatered sludge, averaged per year, in percent.

For economic optimization, the sludge factor S should be as low as possible.

To achieve good performance of the subsequent dewatering and belt drying unit, a maximum alternation of TS of thin sludge needs to be limited to ± 10% relative. This means that the TS of a thin sludge with an average value of, for example, 2% should not fall below 1.8% and not exceed 2.2%. This can be achieved, for instance, with sufficient stirring of thin sludge in a tank with a suitable capacity upstream of the dewatering machine and by avoiding large peak loads in wastewater treatment.

Fig. 4.4 Piston press at HUBER SE laboratory for the testing of achievable dry residue when pressing thin sludge. (HUBER SE 2020)

Measuring the TS of thin sludge fed to the dewatering unit and monitoring polymer dosage are also recommended with regard to economic operation and stable dewatering results.

Laboratory tests for the optimization of sludge dewatering can consist of:

- Using a piston press to optimize polymer selection for maximum dewatering effect, as illustrated in Fig. 4.4.
- Evaluating the DR(A) water content of sludge, which is the maximum achievable DR, corresponding to the maximum free water portion of sludge. This is carried out by gentle drying with a thermo-gravimetric method until a DR marking the end of the constant drying rate is reached. The optimum DR in practical tests with the industrial unit should not fall below this value by more than 2% DR absolute.

The maximum achievable DR (= DR[A]) can be used to predict the maximum DR for industrial dewatering with an accuracy of ± 1.5% DR absolute (Kopp and Dichtl 2001). Thus, according to Table 4.3, adequate dewatering of sludge is reached for different dewatering machines.

Table 4.3 Adequate dewatering of sludge, expressed as the difference of maximum achievable dry residue, DR(A), to achieved DR

Dewatering machine	DR(A)–DR in % absolute
Centrifuge	0%
Belt press/screw press	≤ 2%
Chamber filter press	0%

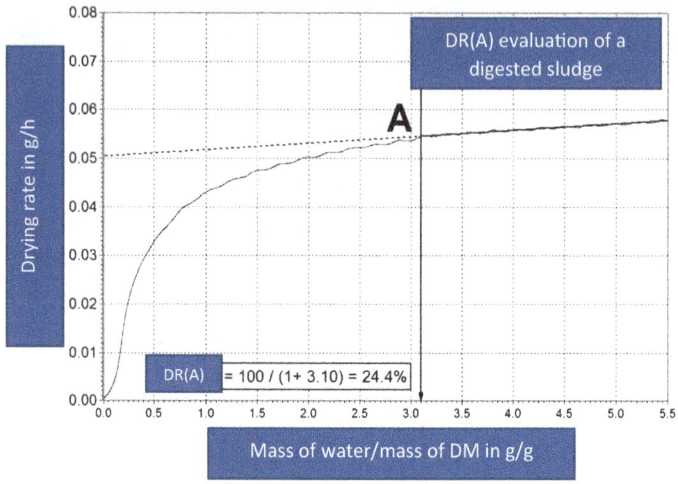

Fig. 4.5 Determination of achievable dry residue, *DR(A)*, by drying. (Kopp 2016)

A DR(A) analysis was carried out for a wastewater treatment plant of 68,000 P (design, 86,500 P) and wastewater flow of 83,660 m^3/day (18.4 millions of gallons per day [MGD]) (City of Sheboygan 2020). Biological phosphorus (Bio-P) elimination was performed, as well as digestion with an additional co-substrate feed of 227 m^3/day (60,000 gallons per day [GPD]). Digestion was carried out in a two-stage process with a total retention time of 40 days and a digester volume of 11,350 m^3 (3,000,000 gallons) (City of Sheboygan 2020). A drying curve of a thin sludge sample like in Fig. 4.5 is monitored with great accuracy, and the point A of expressed increasing of slope of drying curve represents the DR(A). This curve belongs to the digested sludge discharged from the digestion process with extended use of co-substrates from the food industry. The sample was taken in March, i.e., winter sludge was tested. The thin sludge had a TS content of 5.94% and a VSS content of 65.1%. Electrical conductivity was around 9.9 mS/cm, slightly elevated for a digested sludge. Many colloids could be detected between the sludge flocs, filling the free water areas. As a result, the polymer demand for dewatering was elevated. For the liquid polymer, Solenis Praestol K274 (46% of active substance, very high cationic charge), consumption was 11.5 kg active substance per metric ton of DM. The solution concentration was 0.2 mass% for the active substance. The DR(A) value was determined to be 24.4%, and the screw press dewatered the sludge to 22.5%, thus meeting the lower limit for sufficient dewatering of this sludge (Kopp 2016).

Well known consulting experts in sludge dewatering include Prof. Matt Higgens in the US and Dr. Julia Kopp in Germany.

References

City of Sheboygan (2020) Regional Wastewater Treatment Facility. https://sheboygandpw.com/perch/resources/admin/plant-brochure-2020-final-version.pdf. Accessed 19-04-2022

Deutsche Vereinigung für Wasserwirtschaft, Abwasser und Abfall e.V. (DWA) (2008) Characteristic Values of Sludge Dewatering (In German). Merkblatt DWA-M 383. DWA, Hennef

Higgins MJ (2016) Effects of Soluble P on Bioflocs and Dewatering. Sludge Tech Conference, University of Surrey, 26.06.–28.06.2016

HUBER SE (2018) Practical Hints for the Determination of Overdosing of Polymer in dewatering of sludge. Harald Neumann (HUBER SE Germany)

HUBER SE (2020) Testing piston dewatering unit in lab of HUBER SE

Kim YH (1995) Coagulants and Flocculants—Theory and Practice. Tall Oaks Publishing, Littleton

Kopp (2016) Kläranlagenberatung Kopp: Report on Dewaterability of a Sewage Sludge from an US wastewater treatment plant

Kopp J (2002) Water Binding in Sludge and Conditioning (in German). Müllhandbuch Digital Mua Lfg 9/2002:1–18

Kopp J (2014) Parameters of Sludge Dewatering (in German). Training lecture for operating and commissioning installer in Berching. Kläranlagenberatung Kopp, Lengede

Kopp J (2019) Why dewaterability of Sewage Sludge Occurs Upstream. Presentation Biosolids UK.

Kopp J, Dichtl N (2001) Prediction of Full Scale Dewatering of Sewage Sludge by Thermographimetric Measurement of the Water Distribution. Water Sci Tech 42(9):141–149

Kopp J, Ewert E (2006) New Processes for Improvement of Sludge Digestion and Sludge Dewatering. 11th European Biosolids and Organic Resources Conference, Wakefield

López-Paulau S (2015) Sludge preheating system as a strategy to improve sludge dewatering in wastewater treatment plants. Full scale trials. Sludge Tech Conference, University of Surrey, 29.06.–01.07. 2015

SNC Floeger (2003) Publication Sludge Dewatering. Andrézieux Cedex. Fance 01-2003

Stube I, Kopp J (2013) Tendering of Polymer Flocculants—a new approach (In German). Korrespondenz Abwasser Abfal 60(9):780–787

5 Belt Drying of Sewage Sludge

5.1 Systems of Sludge Preparation for Belt Drying

Before dewatered sludge is dried in a belt dryer, the sludge needs to be processed using different techniques:

- Extrusion of sludge by pumping it to an extruder, where strains of 8, 10, or 13 mm are formed and evenly distributed across the first belt. A lower strain diameter of 6 mm is only used for sludge dewatered to more than 26–28% dry residue (DR), thus giving a stable bed of strains. The extrusion process is carried out by pressing sludge through a perforated plate, by forming sludge strains between two rollers, or by the action of a swiveling beam.
- Back-mixing of fully dried sludge to a mixing unit where dewatered and dried sludge are evenly mixed and the DR of the resulting sludge is beyond the sticky phase of the sludge, i.e. > 60–65% DR.
- Tearing the sludge to small pieces by a fast rotating coarse mill and feeding the sludge onto the first belt with the help of a screw conveyor without a bottom shell for even distribution across the belt width.

Fig. 5.1 illustrates an extruder with a perforated plate forming strains. In modern belt drying systems, the feeding device and extruder are covered to minimize air leakage as the dryer is operated under a low vacuum.

As a rule, extrusion quality increases with increasing volatile dry residue (VDR) and decreasing silt content of dewatered sludge. The necessary DR for good extrusion is described with the formula:

$$\mathrm{DR} = \mathrm{DR}_{\max} \times \varepsilon \times a \times \exp\left(\frac{-\mathrm{VDR}}{a}\right) \tag{5.1}$$

DR_{max} is the maximum DR achievable with a piston press in a laboratory (see Chap. 4), ε is the porosity of the sludge strain bed ($\approx$ 0.4–0.5), a is a factor depending on spe-

A. Heindl, *Belt Drying of Sludge*, https://doi.org/10.1007/978-3-662-66512-1_5

Fig. 5.1 Extrusion with a perforated plate. (HUBER 2012)

cific surface of sludge strains ($a = 0.53 + 0.006 \times$ specific surface of strain [in m^2/m^3], $a \approx 1.7$; Schubert 2013). The silt content needs to be less than 70% of the wet grain size distribution analyzed with dewatered sludge.

For endless strains with a circular cross section, the specific surface/volume (S/V) is calculated by:

$$\frac{S}{V} = 3.53 \times (1 - \varphi) \times \frac{1}{d} \tag{5.2}$$

where d: diameter of strain, φ: portion of contact between strains, S: surface in m^2, V: volume in m^3 (Schilp 1977).

The rigidity of the strains is proportional to d^4, while yield stress is proportional to $1/d$. If yield stress of a certain sludge is exceeded, the bed of strains collapses, the contact surface of strains will increase, and drying time will be prolonged. Thus, a larger strain diameter increases the drying time and rigidity of sludge strains. The yield stress, however, decreases and collapsing might occur. On the other hand, smaller strain diameters increase the contact surface and decrease porosity of the bed. With rising VDR of dewatered sludge, the rigidity of the sludge strain decreases as organic sludge particles usually have smaller particle diameters (Luggen et al. 1976). Therefore, the strain diameter needs to be adapted to the rheological structure of dewatered sludge, which is determined by the silt content with a particle diameter of inorganic substances of less than 63 µm, volatile dry mass content with particle diameters down to 5 µm, content of free water, and content of special constituents such as protein. A stable bed with sufficient porosity is necessary for a fast and even drying process.

Fig. 5.2 shows a counter rotating groove roller forming short strains. The dewatered sludge is stored in a hopper, distributed by a scrapper bar system across the belt width, and conveyed to the extruder by a feeder chain (SEVAR 2022). The swiveling beam extruder in Fig. 5.3 operates along the whole width of the belt dryer. A continuously operating

BT
Bandtrockner
Belt Dryer

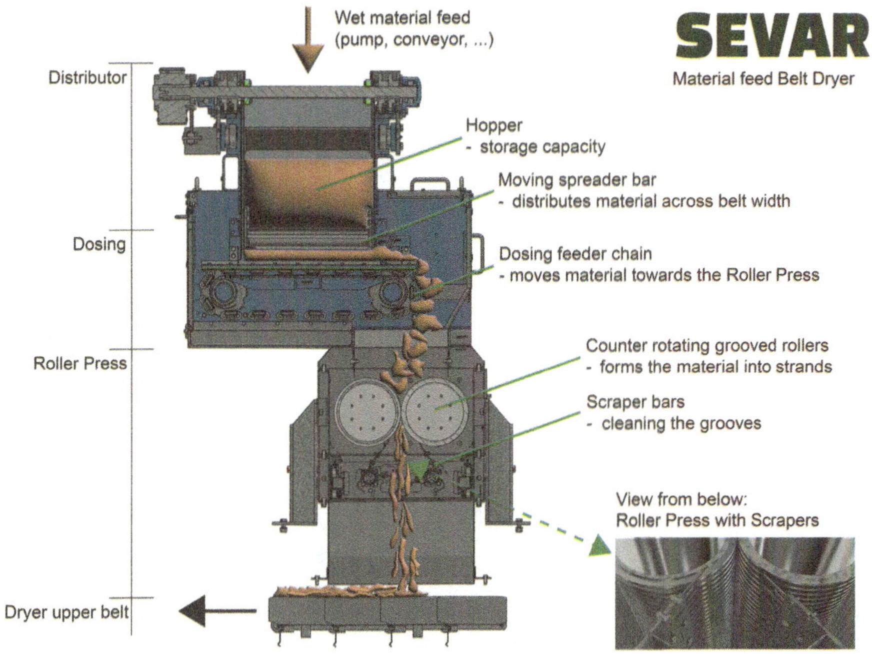

Fig. 5.2 Extrusion with counter-rotating groove rollers and scraper bars cleaning the grooves. (SEVAR 2022)

Fig. 5.3 Swiveling beam extruder. (Klärschlammverwertung Albstadt 2012)

Fig. 5.4 Swiveling perforated-plate extruder for continuous operation. (Heindl 2022)

Fig. 5.5 Dewatered sludge extruded through a perforated plate. (Heindl 2022)

swiveling extruder with a perforated plate and integrated water cleaning loads the strains in Fig. 5.4 onto the first belt. Properly extruded sludge strains resulting from the action of an extruder, as pictured in Fig. 5.1, are shown in Fig. 5.5.

In addition to round holes in the die used for extruding sludge, different shapes such as crossed or striped have been tested in the past. According to theory, higher surface to volume ratios of strains are achieved; their successful application strongly depends on the structural strength of the strains under the influence of free water content, gravity, and temperature.

The sludge tearing option is only feasible for readily digested sludge, which can be transformed to a bulk material suitable to be handled with the screw conveyor at the dryer entrance.

The back-mixing process is almost independent of the sludge structure, but a sophisticated process adjustment in the case of varying properties of dewatered sludge and liability to fiber agglomeration in the mixing device need to be considered. The back-mixing ratio r, i.e., mass of fully dried sludge to mass of dewatered sludge and the mass of mixed sludge fed to the dryer is calculated by (DWA 2021):

$$r = \frac{M_{\text{dried}}}{M_{\text{dewatered}}} = \frac{1 - \frac{\text{DR}_{\text{dewatered}}}{\text{DR}_{\text{mixed}}}}{\frac{\text{DR}_{\text{dried}}}{\text{DR}_{\text{mixed}}} - 1} \tag{5.3}$$

$$M_{\text{mixed}} = M_{\text{dewatered}} \times (1 + r) \tag{5.4}$$

where M: mass of sludge in kg; dewatered: sludge with DR after the dewatering process; dried: sludge with DR $> 90\%$ (fully dried); mixed: sludge resulting from the mixing process of dewatered sludge with dried sludge with a DR_{mixed} of typically 60–65%

As a rule, approximately two-thirds of the dried sludge leaving the dryer is recirculated.

The extrusion process produces sludge strings promoting even drying with minimized dust generation during drying. However, the sludge needs to be dewatered to a DR so that the amount of free water in sludge is low, i.e., less than 2% DR (absolute) below the maximum possible DR(A) for dewatering to form a stable bed of strains.

For all systems, a good and even distribution of sludge across the width of the first belt is essential for operating the belt dryer at its designed evaporation level and obtaining a DR of dried sludge $> 90\%$ with small deviations.

5.2 Specific Water Evaporation

The belt dryer is designed according to the necessary water evaporation and not according to the throughput of dewatered sludge.

Specific water evaporation (SWE) in belt dryers is determined by various factors:

- Air temperature: Typically air temperatures range from 50 to 100 °C (122 to 212 °F) and at maximum 125–130 °C (257 to 266 °F).
- Air speed through the bed of sludge: Typically, air speed ranges from 0.4 to 1.2 m/s (78.6–236.4 ft/min), depending on the type of belt drying system.
- Degree of air recirculation influencing absolute humidity of drying air: Typically, the recirculation rate is approximately 80–90% of total air flow.
- Condensing system or fresh air–exhaust air system without condensing.
- Sludge constituents (protein, fat) and type as well as degree of sludge stabilization.
- Preparation of dewatered sludge (extrusion, back-mixing).
- Surface to volume ratio of sludge strains/granules fed to the dryer (extruding diameter): Larger strain diameters increase the stability of the bed but decrease the surface to volume ratio. Endless strains with a diameter of 10 mm and a contacting ratio of 0.15

have a theoretical specific surface of $330\,m^2/m^3$, whereas for 6 mm strains, the surface to volume ratio is around $550\,m^2/m^3$ (Poxleitner 2012).

Belt dryers without condensing systems operate at an absolute humidity of drying air at the dryer inlet of approximately 6–12 g of water vapor per kg of dry air, depending on the conditions of ambient air. Condensing systems, however, have a higher absolute humidity of air recirculated to the dryer in the range of 30–50 g/kg of dry air. Thus, the capacity of water vapor uptake is higher in the case of fresh air–exhaust air systems, resulting in a higher specific evaporation rate and lower demand in terms of drying area. However, the exhaust airflow of the latter system is considerably higher in comparison to the condensing system.

For digested sludge, tests showed an exponential dependency of drying time to a DR > 90% from air temperature, extrusion diameter, and air speed (assuming equal porosity and height of the sludge bed; Poxleitner 2012):

$$\text{Air temperature: change in drying time} \approx (\vartheta/\vartheta_0)^{-1.7} \tag{5.5}$$

$$\text{Extrusion diameter: change in drying time} \approx (d/d_0)^{0.162} \tag{5.6}$$

$$\text{Air speed: change in drying time} \approx (w/w_0)^{-0.55} \tag{5.7}$$

The index 0 means a known value. Thus, for example, increasing air temperature from 80 to 100 °C would reduce drying time to 68.4%—or increasing air speed from 0.9 to 1.1 m/s would result in a corresponding drying time of 89.6%—of the initial value. To conclude, the air temperature has the most pronounced effect on drying time and consequently on water evaporation and dryer size.

As a rule of thumb, specific water evaporation at an air temperature of 125–130 °C ranges from 22 to 28 kg $H_2O/(m^2$ of belt drying area × h) (4.51 to 5.73 lbs/ft^2), showing the influence of sludge type, sludge constituents, and degree of stabilization. Low-temperature sludge drying at an air temperature of around 48–50 °C results in a specific water evaporation at approximately 6 to 7 $kg/(m^2 \times h)$ (1.23 to 1.43 lbs/ft^2). SWE values for air temperatures of around 80–95 °C are typically encountered at between 13 and 17 $kg/(m^2 \times h)$ (between 2.66 and 3.48 lbs/ft^2). Another source indicates that the relative water evaporation with air temperatures of 80 °C:100 °C:130 °C is 1:1.5:1.8 for the drying of sludge in a two-belt dryer with extrusion (Haarslev Industries 2016). Thus, this clearly shows the considerable effect of air temperature on drying rate and consequently on dryer size for a given water evaporation. The specific water evaporation capacity of a condensing system is slightly lower in comparison to fresh air–exhaust air drying systems, since recirculated air usually has a higher absolute water content after dehumidification in the condenser (≈ 30–50 g/kg of dry air) in comparison to ambient air (≈ 6–12 g/kg of dry air). Therefore, the fresh air–exhaust air dryer shows a higher capacity of water vapor uptake of drying air.

Especially sludge with high protein values and high content of lipophilic substances tends to form case hardening on the surface of sludge particles or extruded strains. The outer surface has a higher concentration of these substances and a barrier is created when the surface dries out. This effect retards water movement from the kernel to the surface of sludge particles, and consequently water evaporation as well as throughput of the dryer decrease considerably. Similar effects can be encountered with industrial, biological sludge from the paper industry where polysaccharides and fibers may cause case hardening and plate formation in dryers. Case hardening effects can reduce SWE in extreme cases to 75 or even 50% of the values normally encountered with well-stabilized sludge.

5.3 Belt Drying Systems

Nowadays, various belt drying systems are in operation all around the world. These systems are characterized by the following main features:

- Number of belts: Single-stage belt dryer (one-belt dryer) or multi-stage belt dyer (two-, three-, or four-belt dryers).
- Preparation of sludge fed to the dryer: Extrusion or back-mixing systems.
- Exhaust air treatment: Condensing or non-condensing fresh air–exhaust air systems.
- Direction of air flow: Air flow upwards or downwards through the sludge layer on belt(s).
- Size of belt dryers: Sizes range from small units installed in containers of 20 m^2 drying areas to very large belt dryers for outdoor installation of up to 400 m^2.

Upward air flow is adapted to the extruded sludge and has the advantage of lower air resistance. However, air speed is limited by the lifting speed of the sludge on the belts. Downward air flow is used for back-mixed sludge as well as for extruded sludge and has the advantage of the filtering effect of the sludge layer. In this case, air speed is limited by increased resistance of the compressed sludge layers, and sludge dust deposits often form at the base of the dryer. For single-belt dryers with back-mixing, downward air flow speed is limited by approximately 0.4–0.5 m/s ($\approx$ 80–100 ft/min), while belt dryers with sludge extrusion allow air speeds of up to 1.0–1.2 m/s ($\approx$ 200–236 ft/min) for air-up or air-down flow.

5.4 Belt Dryers for Sewage Sludge

Belt dryers are continuously operating drying systems. For economic, energy, and operational reasons, they need to be run continuously 24/7 for up to 8000 h per year. This ensures continuous, trouble-free operation, long-term operation, and a consistent final dry residue of fully dried granules. Frequent start-ups and shut-downs of the dryer may result in higher energy consumption, higher fluctuations in the DR of dried material, condensation and corrosion problems, as well as operational issues in sludge feeding and discharging.

5.4.1 Systems with Extrusion of Dewatered Sludge

The HUBER System applies extrusion of sludge and drying in a two-belt dryer with direct or indirect helical airflow and indirect sectional heating of circulated drying air. Fig. 5.6 presents a dryer of this kind at the wastewater treatment plant of Innsbruck, Austria, which is in operation since 2019.

It has a sectional design with an individual circulating fan and hot water air heater for each section. The air is heated by horizontally arranged hot-water–air-heat exchangers in each section before flowing through the bed of sludge strains. Heated air transfers energy to the sludge, takes up water vapor, and is cooled before it is transferred to the next section by the circulating fan. Thus, drying air and sludge are conducted in cross-current flow for the first belt and in counter-current cross flow for the second belt. Air leaving the dryer with a high water vapor content is conducted through a condensing line consisting of indirect air heaters at the beginning and at the end of the line, forming an internal heat recovery system. In an intermediate heat exchanging system—either indirect with finned tubes or direct with sprayed packed bed—water vapor is condensed and air is dehumidified before it is reheated and circulated back to the dryer. A small proportion of the air is conducted to the exhaust air treatment system as shown in Fig. 5.7.

Fig. 5.6 Two-belt dryer at the wastewater treatment plant in Innsbruck, Austria; the transfer station from belt one to belt two is shown, together with the high pressure pump for belt cleaning

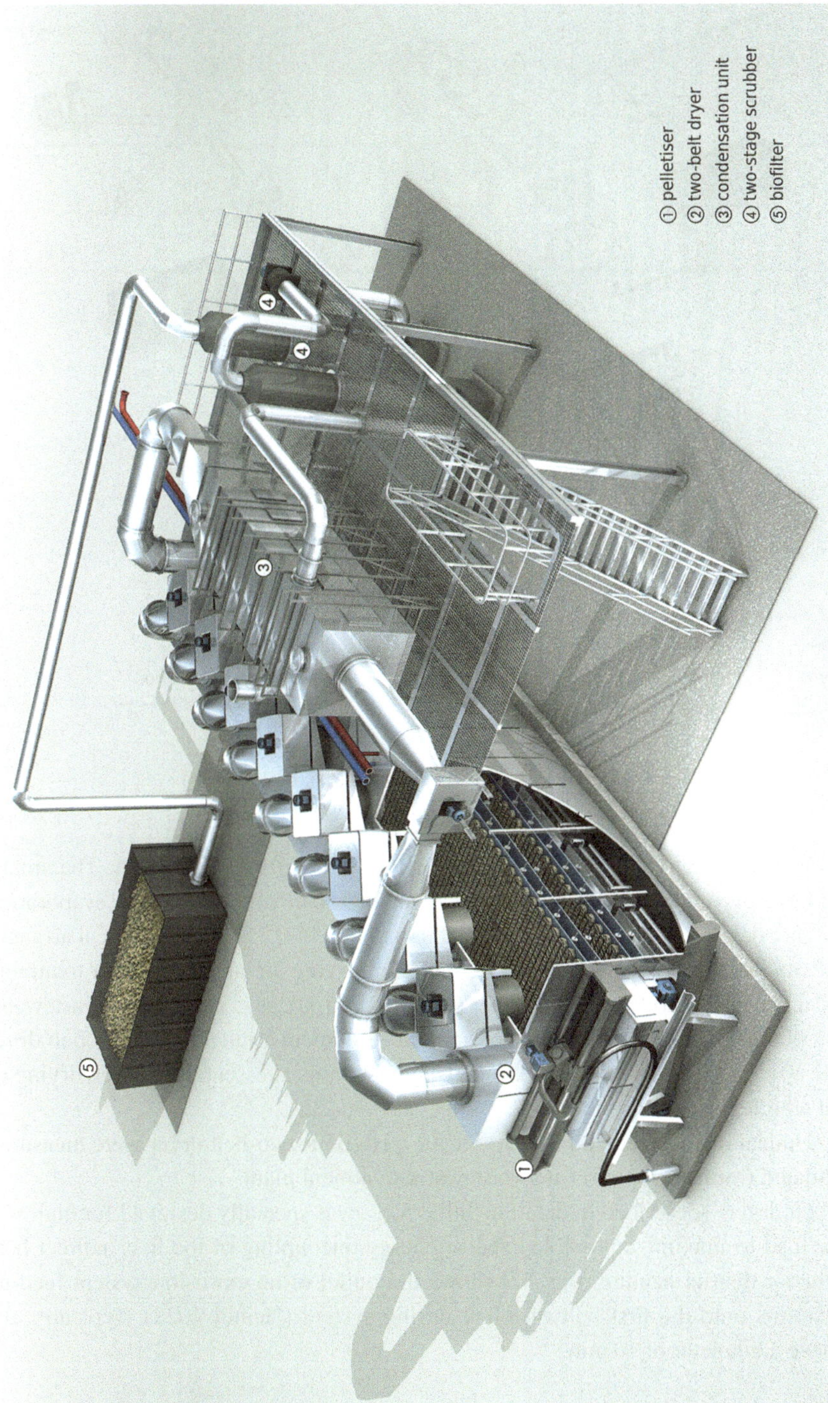

Fig. 5.7 General arrangement of a two-belt dryer with condensing line and exhaust air treatment. (HUBER 2016a)

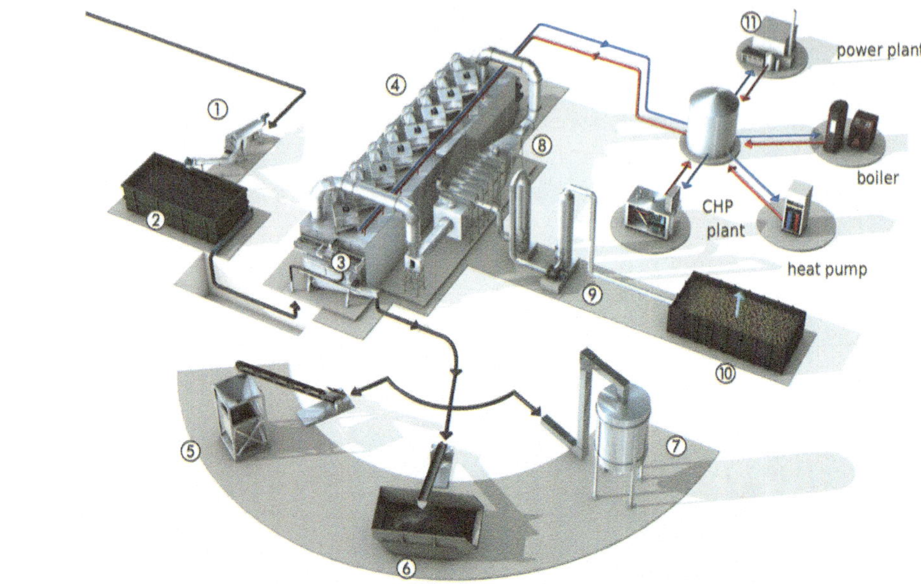

① screw press for sludge dewatering
② sludge bunker with thick sludge pumps
③ pelletiser
④ Belt Dryer
⑤⑥⑦ granulate discharge to big bags, open containers or silos
⑧ condensation unit
⑨ exhaust air scrubber
⑩ biofilter for exhaust air
⑪ wastewater heat utilisation (power plant, boiler, heat pump, CHP plant)

Fig. 5.8 Peripheral equipment arranged around a belt dryer. (HUBER 2016b)

The standard effective belt width is 2.5 m, and the module length is 2 m. The smallest dryer has a drying area of 20 m^2, the tallest one of 150 m^2. If higher water evaporation capacities are desired, several lines are placed in parallel. Fig. 5.7 shows a typical arrangement with an indirect condensing line, circulation of drying air, and exhaust air treatment with acid and caustic scrubber and biofilter (HUBER 2016a), as is common on wastewater treatment plants. Fig. 5.8 illustrates the peripheral equipment arranged around a belt dryer (HUBER 2016b). Good stability of extruded sludge strains is essential for even drying of sludge at a high performance level.

The technical data presented in Table 5.1 for a HUBER two-belt dryer were measured and calculated from a test period at a wastewater treatment plant.

This belt dryer—described in detail in Table 5.1—was specially designed for high water vapor load to maximize possible exhaust energy uncoupling of the first indirect heat exchanger for district heating. Fig. 5.9 shows the outlet of an extrusion system feeding endless strains onto the first belt of a belt drying system (Heindl 2022). Typically, the strains have a diameter of 10 mm.

Table 5.1 Technical test data of a two-belt-dryer with extrusion of dewatered, digested sludge. (HUBER 2019a)

Type of belt dryer	Two-stage with extrusion of dewatered sludge and two drying zones, indirect helical air flow
	Condensing system with external heat for district heating network
Sludge	Digested, biowaste as co-substrate (primary sludge ≈ 34%, secondary sludge ≈ 43%, biowaste sludge ≈ 23%, yielding 15,000 metric tons of dewatered sludge with an average DR of 24%)
Throughput of dry mass (DM)	≈ 670 kg/h
Dry residue (DR) of dewatered sludge	≈ 24%
Feeding of dewatered sludge	≈ 2730 kg/h
Water evaporation for drying to 90% DR	≈ 2000 kg/h
Drying time	≈ 100 min
Air temperatures	85 / 125 °C in zone 1 (first half of first belt and second half of second belt) / 2 (second half of first belt and first half of second belt)
Drying area	≈ 80 m^2
Feeding height of dewatered sludge	≈ 10 cm
Specific water evaporation (SWE)	≈ 25 $kg/(m^2 \times h)$
Exhaust energy warm water of CHP (combined heat and power system)	≈ 410 kW
Specific thermal energy consumption	0.80–0.81 kWh/kg water evaporation
Specific electric energy consumption	0.037–0.040 kWh/kg water evaporation
Exhaust air flow	≈ 4000 m^3/h
Exhaust air treatment	Two-stage scrubber and biofilter
Thermal energy for district heating, recovered from exhaust air of dryer in front of direct condensing unit	≈ 400 kWh/h at 70 °C
Exhaust air treatment	Acid and caustic scrubber, biofilter

The SÜLZLE-KLEIN system also operates with the extrusion of sludge and drying in a two- or three-belt dryer with sectional circulation of air (SÜLZLE-KLEIN 2022). Fig. 5.10 shows a two-belt dryer with a sectional arrangement of gas-fired chambers for direct heating of recirculated air. The large yellow pipe is designed for the overall gas supply of the three gas burners. Direct firing has the advantage of a higher thermal efficiency, but attention needs to be paid to sufficient mixing of flue gas with recirculated air to avoid air strains with different temperature ranges, and dust deposits need to be removed in a timely manner. In addition, only natural gas with close to zero sulfur content is recommended in order to avoid corrosion issues in the case of condensing.

Air heaters are arranged in a vertical position, and recirculated and reheated air flow is directed downwards through the layer of extruded sludge on the belts. Table 5.2 shows

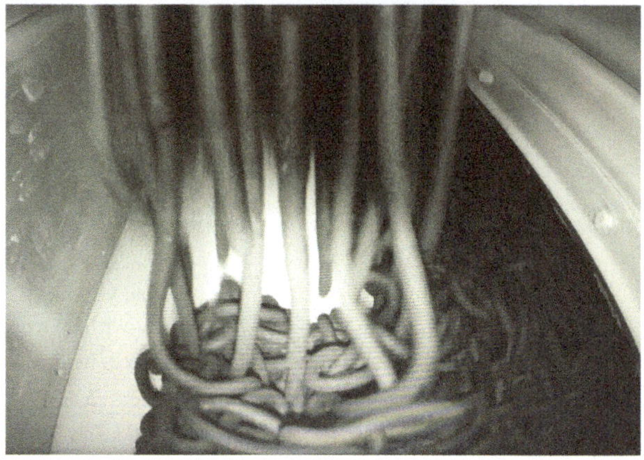

Fig. 5.9 Extrusion system of a HUBER belt dryer feeding endless strains onto the first belt. (Heindl 2022)

the data of a two-belt dryer with extrusion system, mentioned in a company publication (SÜLZLE-KLEIN 2016).

Fig. 5.11 illustrates a two-belt dryer with steam heating of air; the large insulated pipe at the top is for the steam supply, while the large insulated pipe on the bottom is to discharge the condensate.

The STELA system runs with extrusion of sludge or partial back-mixing of dried sludge and drying in a one- or two-belt dryer with partial and sectional recirculation of

Fig. 5.10 SÜLZLE-KLEIN two-belt dryer with gas-fired chambers for sectional direct heating of recirculated air

Table 5.2 Technical data of a two-belt dryer for digested sludge with an extrusion system

Type of belt dryer	Two-belt with extrusion of dewatered sludge and two temperature zones
	Sectional recirculation of air in four zones
Sludge	Digested municipal sludge
Throughput of dry mass (DM)	$\approx 534\,kg/h$
Dry residue (DR) of dewatered sludge	$\approx 29\%$
Feeding of dewatered sludge	$\approx 1840\,kg/h$
Water evaporation for drying to 90% DR	$\approx 1306\,kg/h$
DR of dried sludge	$> 90\%$
Heating of drying air	Indirectly via tube heat exchanger
Water temperatures zone 1/2	90/140 °C
Air temperatures zone 1/2	$\approx$ 80/125 °C
Air flow	Not indicated
Drying area	$\approx 80\,m^2$
Specific water evaporation (SWE)	$\approx 20.4\,kg/(m^2 \times h)$
Specific thermal energy consumption	0.80 kWh/kg water evaporation
Specific electric energy consumption	$<$ 0.096 kWh/kg water evaporation
Exhaust air flow	$< 7500\,m^3/h$ at 40 °C
Exhaust air treatment	Two-stage scrubber (acid and caustic) and biofilter

Fig. 5.11 Two-belt dryer for sludge with steam heating. (Heindl 2022)

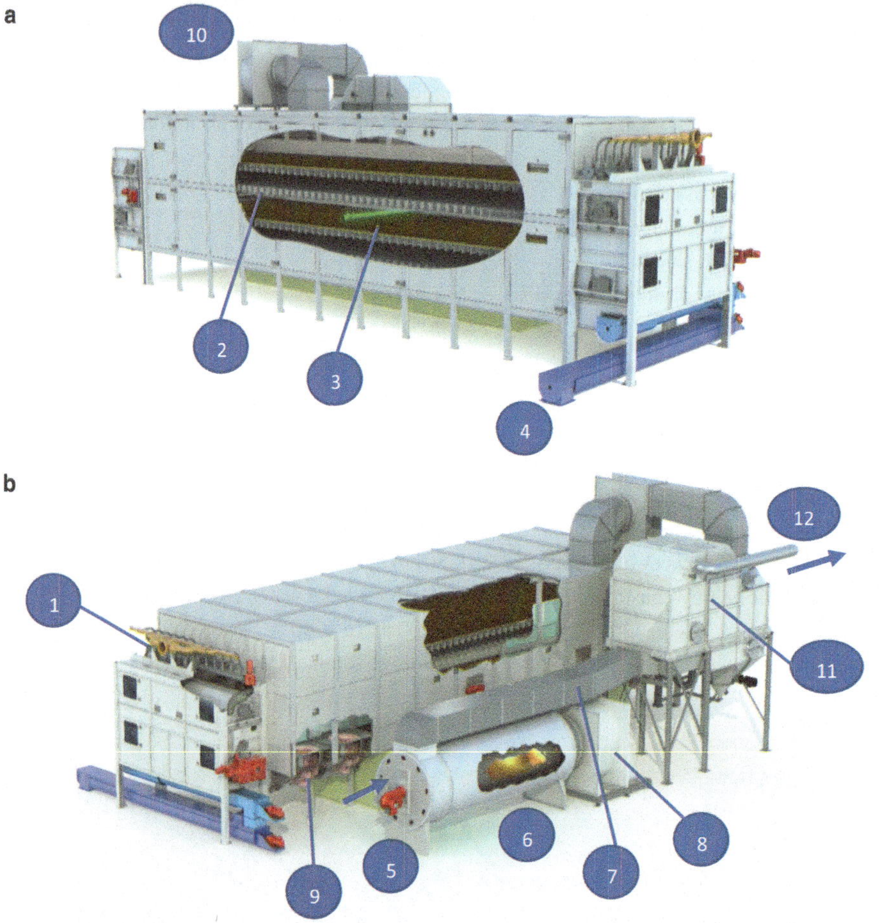

Fig. 5.12 **a,b** STELA two-belt dryer with extrusion system and direct-fired heating of air. (STELA 2023). *1* Extrusion unit; *2* perforated steel plate belt; *3* sludge stirring equipment situated at half-way point of second belt; *4* water cooled discharging screw; *5* fresh air intake at gas burner plate; *6* hot air generator; *7* exhaust air tube for heat recovery; *8* hot air ventilator; *9* sectional circulating ventilators; *10* exhaust air ventilator for heat recovery; *11* exhaust air heat exchanger (cooling and condensing); *12* exhaust air

air. Fig. 5.12a, b shows a two-belt dryer with an extrusion system and direct heating of drying air.

The extrusion unit consists of a system of pipes feeding a final pipe running across the width of the belt. This final pipe releases sludge onto the belt via a number of holes, while inside the pipe a rotating cleaning device is continuously removing hairs of fibers from holes.

Fig. 5.13 Two-belt dryer with a group of feeding lances (*left side top* with flexible tubes for sludge feeding) and a scrubber for exhaust air cleaning. (Veolia 2022a)

The belt dryers are designed for a drying area of up to 380 m^2. The temperature of exhaust energy sources is typically up to 140 °C for woven plastic belts.

The VEOLIA system comprises extrusion of sludge and drying in a two-belt dryer with partial recirculation of air. The transfer DR from belt 1 to belt 2 arranged in a line is approximately 50%. Dewatered sludge is fed to the dryer belt with several groups of swiveling hollow lances, each ending with a nozzle. Each lance group has its own sludge pump. Fig. 5.13 illustrates the arrangement of this dryer (Veolia 2022a).

Fig. 5.14 shows the air flow in the dryer. Air temperatures for the first belt are set to 120–170 °C and for the second belt to 80–100 °C. With this temperature regime, the drying time of extruded sludge is indicated to be around 60 min.

Table 5.3 shows data of a two-stage one-belt dryer with an extrusion system, mentioned in a company publication (Veolia Krüger 2012).

The BINDER system is a one-belt belt dryer with a swiveling extruder (Heindl 2022).

This single-stage belt dryer consists of several modules in serial arrangement and sealed off from each other. Fig. 5.15 highlights the general arrangement of such a dryer (Heindl 2022). Each module is equipped with one circulating fan and one air heater. Thus, a reversion of flow of heated air is possible without cross flow effects between the modules to achieve more even drying of sludge strains on the belt. The fans are integrated into a dryer casing and can be swept out for maintenance purposes. The belt is designed as a chain conducted belt with hinged perforated metal plates. The belt is cleaned with a rotating brush at the discharge end of the dryer by removing dried sludge particles. Dewatered sludge is extruded by a swiveling headpiece with a perforated dye, as shown in Fig. 5.4. This dye is

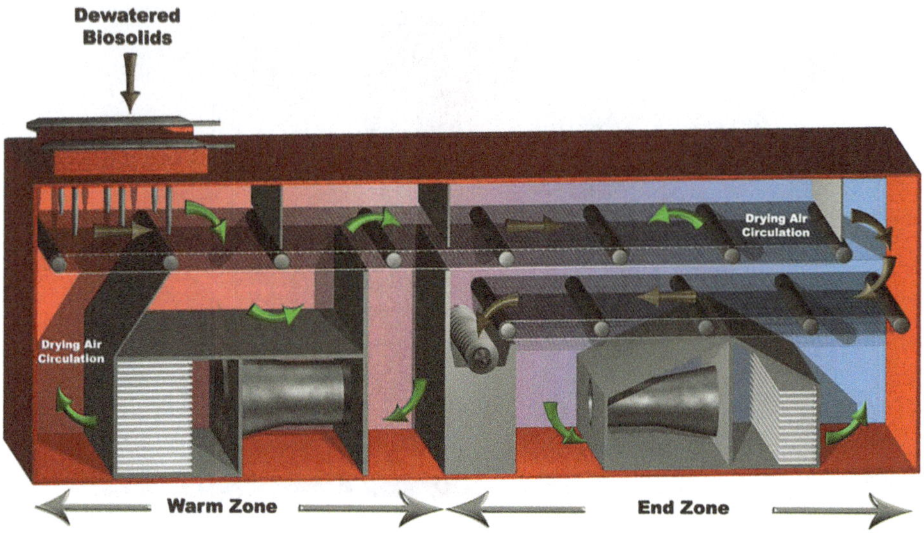

Fig. 5.14 Sectional view of the Bioco two-belt dryer. (Veolia 2022b)

Table 5.3 VEOLIA two-stage one-belt dryer

Type of belt dryer	Two-stage one-belt dryer for aerobically stabilized sludge and three temperature zones Sectional recirculation of air in three zones
Sludge	Mixed sludge of primary and biological sludge
Throughput of dry mass (DM)	≈ 190 kg/h
Dry residue (DR) of dewatered sludge	≈ 19%
Feeding of dewatered sludge	≈ 1000 kg/h
Water evaporation for drying to 90% DR	≈ 789 kg/h
DR of dried sludge	> 90%
Heating of drying air	Indirectly via tube heat exchanger
Thermal oil and water temperatures zone 1/2/3	220/95/85 °C
Air temperatures zone 1/2/3	≈ 160/85/75 °C
Air flow	Not indicated
Drying area	≈ 40 m^2
Specific water evaporation (SWE)	≈ 19.7 $kg/(m^2 \times h)$
Drying time	≈ 60 min
Specific thermal energy consumption	Not indicated
Specific electric energy consumption	Not indicated
Exhaust air flow	Not indicated
Exhaust air treatment	Direct scrubber

Fig. 5.15 One-belt dryer with swiveling extruder. (Heindl 2022)

cleaned frequently and automatically with water at a point adjacent to the conveyor belt. In this case, there is continuous, linked operation of the extruder and conveyor belt of the dryer.

The technical data for the operation of such a belt drying system are shown in Table 5.4 (practical operator data)

The SEVAR two-belt dryer is equipped with a special material feeding device with counter-rotating groove rollers and scraper bars to form strains. These strains are evenly distributed on the first conveyor belt. The two belts consist of stainless steel perforated plate conveyors (SEVAR 2022). The belt width is 3 m and the number of modules arranged in line ranges from 4 to 18. Each module has a length of 2 m and a total drying area on two belts of 12 m^2. Thus, the available drying surface for one dryer line ranges from 36 to 216 m^2. The modules are divided into a process chamber and a mixing chamber, as shown in Fig. 5.16 for an indirect heated dryer (SEVAR 2022). The circulating fans are integrated into a dryer casing. Air is looped through the dryer and reheated at each dryer

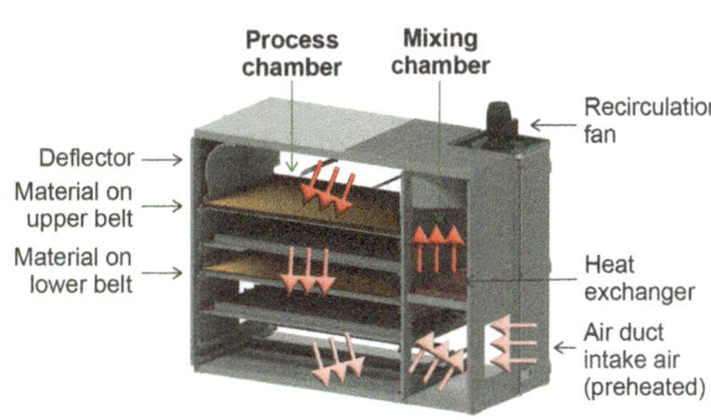

Fig. 5.16 Cross section of indirectly heated belt dryer. (SEVAR 2022)

Table 5.4 Practical operation data of a one-belt dryer with swiveling extruder

Type of belt dryer	Single-stage with extrusion of dewatered sludge (15 mm × 4 mm) and four drying zones
	Fresh air–exhaust air system with internal heat recovery
Sludge	Digested municipal sludge with a high proportion of wastewater from the dairy industry
Throughput of dry mass (DM)	Approximately 138 kg/h
Dry residue (DR) of dewatered sludge	≈ 23%
Feeding of dewatered sludge	≈ 600 kg/h
Water evaporation for drying to 90% DR	≈ 447 kg/h
Drying time	120–130 min
Air temperatures	77/81/88/86 °C in zones 1/2/3/4
Drying area	≈ 27 m^2
Feeding height of dewatered sludge	≈ 10 cm
Specific water evaporation (SWE)	16.6 kg/(m^2 × h)
Exhaust energy warm water of CHP (combined heat and power system)	≈ 410 kW
Specific thermal energy consumption	0.85 kWh/kg water evaporation
Exhaust air flow	≈ 9000 Nm^3/h
Exhaust air treatment	Two-stage neutral scrubber with addition of chemical agent for stabilization of water hardness, metal woven filter, activated carbon filter, UV unit
Exhaust air fan	11 kW
Number of UV lamps	≈ 180, equipped with protective quartz glass cylinders for easy cleaning, protective glass reduces UV radiation by around 15%
UV electrical connection	17.3 kW or ≈ 1.9 kWh/1000 Nm^3
Mass of activated carbon	≈ 6000 kg of ≈ 1.5 Nm^3 exhaust air/(kg of activated carbon × h)
Change of activated carbon	Once per year

section. Air flow through the bed of strains is downwards. The exhaust air flow is around 3.3–3.6 Nm^3 dry/kg of water evaporation.

At hot air temperatures of 120–130 °C, drying time is indicated to be in the range of 50–70 min. During regular shut-down periods, it is recommended to remove fine particles agglomerating on the dryer floor. The proportion of recirculated air is indicated to be more than 90%.

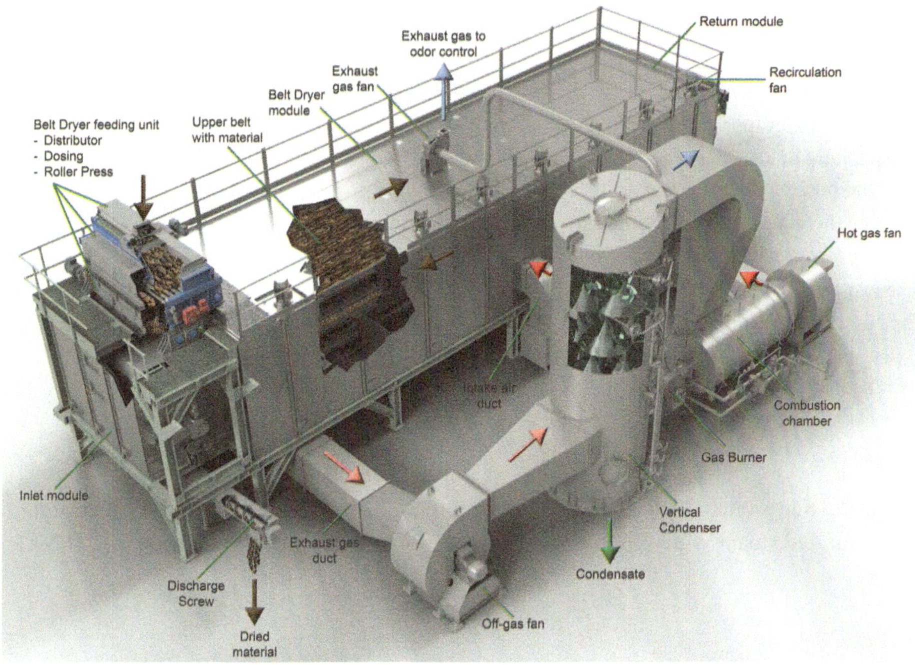

Fig. 5.17 SEVAR two-belt dryer with condensing system and direct heating of recirculated and dehumidified air. (SEVAR 2022)

Fig. 5.17 illustrates the elevation of a two-belt dryer with direct heating of drying air. All elements of the belt drying system are marked in the figure. The direct spraying condenser has special fluid distribution devices to increase the surface of mass transfer. The air flows are marked with arrows.

The data of such a dryer are shown in Table 5.5 (SEVAR 2013).

Table 5.5 Two-belt SEVAR-type dryer according to company data

Type of belt dryer	Two lines of two-belt dryers with extrusion of dewatered sludge
	Direct condensing system
Sludge	Digested municipal sludge
Throughput of dry mass (DM)	$2 \times 362.5\,\text{kg/h}$
Dry residue (DR) of dewatered/dried sludge	25/90%
Feeding of dewatered sludge	$\approx 2 \times 1450 = 2900\,\text{kg/h}$
Water evaporation for drying to 90% DR	$2 \times 1050 = 2100\,\text{kg/h}$
Drying time	Not indicated
Air temperatures	≈ 80–$85\,^\circ\text{C}$
Recirculation rate of dryer air	90–95%

Table 5.5 (Continued)

Drying area	$2 \times 72\,m^2$
Feeding height of dewatered sludge	Not indicated
Specific water evaporation (SWE)	$14.6\,kg/(m^2 \times h)$
Temperature of warm water of CHP (combined heat and power system) in/out	95/75 °C
Specific thermal energy consumption	Not indicated
Specific electric energy consumption	Not indicated
Exhaust air flow	$18{,}000\,Nm^3/h$
Exhaust air temperature dryer	≈ 60 °C
Exhaust air temperature before exhaust air treatment after heat recovery	≈ 45 °C
Exhaust air treatment	Acid scrubber and biofilter

Fig. 5.18 presents a belt dryer with an indirect heating system, supplied for example with saturated steam with a minimum pressure of 4 bar (absolute) or thermal oil (SEVAR 2022).

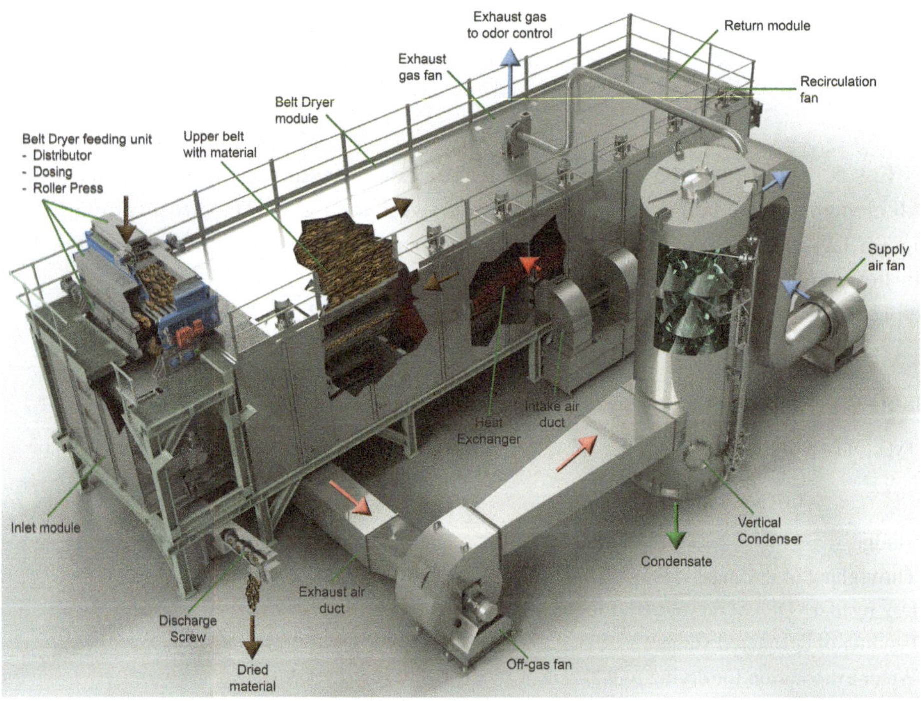

Fig. 5.18 SEVAR two-belt dryer with condensing system and indirect heating of recirculated and dehumidified air. (SEVAR 2022)

Typical thermal energy consumption for these belt dryers at an air temperature of 70–90 °C is indicated to be around 0.9 kWh/kg of water evaporation; at higher air temperatures of 120–140 °C (248–282 °F), the value is around 0.85 kWh/kg of water evaporation according to data furnished by the dryer manufacturer (SEVAR 2022). Heating can be carried out indirectly with horizontally arranged tube heat exchangers or directly with natural gas or oil. The exhaust gas of a combined heat and power system (CHP) used at site can be mixed into the burner plenum to save energy.

5.4.2 Systems with Back-Mixing of Dried Sludge

Drying systems comprising a one-belt-dryer with partial back-mixing of sludge are supplied by different companies, e.g. STELA, but two-belt dryers are also used. The system is described in Fig. 5.19. For one-belt dryers, the feeding height of the mixed sludge ranges from a minimum of 4 to a maximum of 20 cm, while it is typically around 7 cm.

Standard-design dryers with a back-mixing system using direct heating of drying air are equipped with the following devices: dewatered sludge bunker; mixing device for dewatered sludge and a portion of fully dried and sieved sludge (fines); sieving screw conveyor and normal screw conveyor securing a separation of dried sludge into fines to be recirculated and larger particles to be transported to the storage unit; conveyor band for cooling of larger particles; fresh air ventilator; direct fired air heater for recirculated air; heat exchanger for heat recovery from external exhaust energy; belt dryer; recirculating ventilator; exhaust air ventilator; direct spraying condenser for dehumidifying of circulated air and further exhaust air treatment, consisting of chemical scrubbers and biofilter.

Fig. 5.20 shows an elevation of a single-belt dryer with partial back-mixing of sludge and indirect heating of air (STELA 2023). Air flow is directed vertically from top to bottom through the bed of mixed and granulated sludge, which is fed across the belt width by

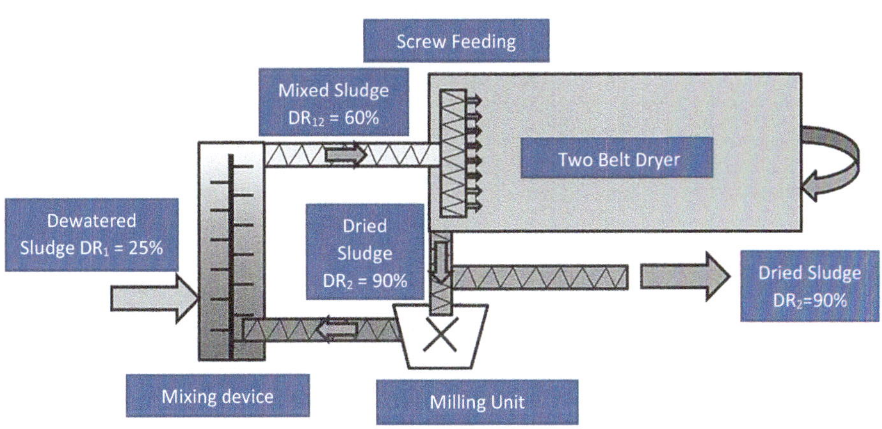

Fig. 5.19 Two-belt dryer with partial back-mixing of dried sludge. (Heindl 2015)

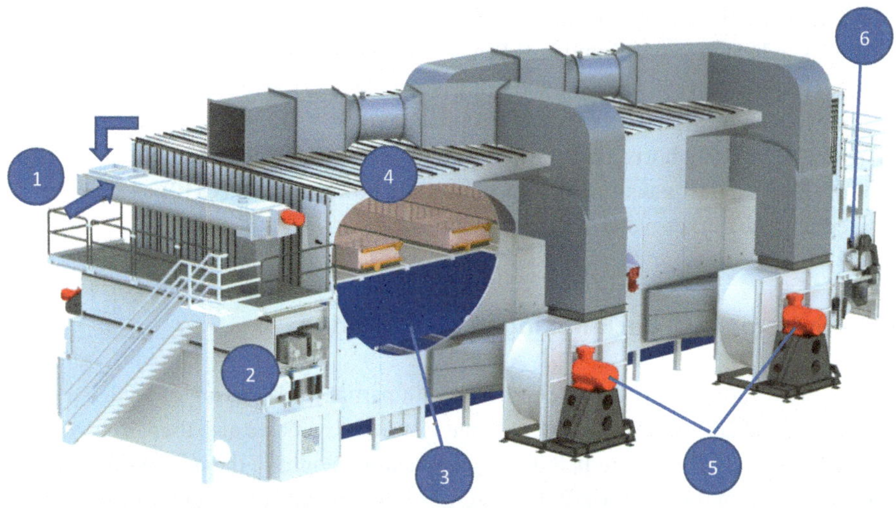

Fig. 5.20 Single-belt dryer with partial back-mixing of dried sludge. (STELA 2023). *1* Mixing screw conveyor fed with fully dried sludge and dewatered sludge; *2* double-dosing screw conveyor, adjustable in height for dosing of mixed sludge in a layer of a certain thickness across the belt width; *3* woven plastic belt; *4* heat exchangers; *5* fans for air circulation and exhaust air; *6* discharging screw conveyor for dried sludge

a double screw conveyor. A portion of dried sludge is transferred back to the feeding end of the dryer and mixed with dewatered sludge in a mixing screw running across the width of the dryer.

Technical data of operation of such a belt drying system—with partial back-mixing of dried sludge—are shown in Table 5.6 (operator data and amendments/own calculations).

Table 5.6 Single-belt dryer for sewage sludge with partial back-mixing of dried sludge

Type of belt dryer	Single-stage with back-mixing of dried sludge and four drying zones
	Fresh air–exhaust air system with internal heat recovery
Source of sludge	With intermediate storage, municipal sludge (digested—aerobically stabilized)
Throughput of dry mass (DM)	$\approx 3800\,\mathrm{kg/h}$
Dry residue (DR) of dewatered sludge	$\approx 25\%$
Feeding of dewatered sludge	$\approx 7600\,\mathrm{kg/h}$
Recirculation of dried sludge	$\approx 1400\,\mathrm{kg/h}$
Water evaporation for drying to 90% DR	$\approx 5500\,\mathrm{kg/h}$
DR of dried sludge	$> 90\%$

Table 5.6 (Continued)

DR of mixed sludge entering dryer	≈ 65%
Heating of drying air	Indirectly via tube heat exchanger
Air temperatures	110–115 °C
Air flow	≈ 320,000 m^3/h
Drying area	≈ 220 m^2
Drying time	≈ 60 min
Feeding height of dewatered sludge	≈ 6 cm
Specific water evaporation (SWE)	≈ 25.0 $kg/(m^2 \times h)$
Specific thermal energy consumption	Not indicated
Exhaust air flow	≈ 60,000 $Nm^3{}_{dry}/h$
Exhaust air treatment	Two-stage scrubber (acid and caustic) and activated carbon filter

Technical data for the operation of a further belt drying system with partial recirculation of dried sludge are given in Table 5.7 (GKD 2021).

It is important to note that the high SWE value is due to the high air temperature of 140 °C in the feeding half of the dryer based on an energy supply via water vapor of 8 bar absolute and 200 °C. The same situation is encountered in the Table 5.8, describing the conditions of a belt drying system in northern Italy and indicating the measured data of the local operator.

Table 5.7 Technical data of a single-belt dryer with partial back-mixing of dried sludge (GKD 2021)

Type of belt dryer	Single-stage with partial recirculation of dried sludge with tow temperature zones
Thermal energy consumption	13,000 MWh/y
Capacity	21,600 t of dewatered sludge per year
	2700 kg/h for 8000 h/y
Dry residue (DR) of dewatered sludge	22% average
Feeding height of mixed sludge	80 mm
Production of dried sludge	5280 t per year
DR of mixed sludge entering dryer	≈ 65%
DR of dried sludge	> 90%
Water evaporation	≈ 2040 kg/h
Drying time	30–50 min
Air temperatures	140/100 °C feeding to discharging
Drying area	≈ 63 m^2
Specific water evaporation	32.4 $kg/(m^2 \times h)$
Specific thermal energy demand	0.80 kWh/kg of water evaporated
Sources of sludge	External from 40 different wastewater treatment plants

Table 5.8 Belt dryer of a wastewater treatment plant (WWTP) in northern Italy. (Engl 2010)

Type of belt dryer	Single-stage with partial recirculation of dried sludge with low temperature zones
Thermal energy consumption	52,000 MWh/y
Capacity	≈ 22,000 t of dewatered sludge per year 2775 kg/h for 8000 h/y
Source of sludge	Sludge of WWTP
Dry residue (DR) of dewatered sludge	20.6% average
Feeding height of mixed sludge	70 mm
Production of dried sludge	4888 t per year
DR of mixed sludge entering dryer	≈ 60%
DR of dried sludge	93.5%
Water evaporation	≈ 2164 kg/h
Thermal energy demand	≈ 1800 kW
Drying time	Not indicated
Air temperature in zones 1/2/3	150/137/110 °C
Drying area	≈ 61 m^2 (3.6 m × 17 m)
Dryer air flow	≈ 134,000 m^3/h at 72 °C
Specific water evaporation	≈ 35.4 kg/(m^2 × h)
Specific thermal energy demand	0.83 kWh/kg of water evaporated
Specific electric energy demand	0.064 kWh/kg of water evaporated
Exhaust air flow	≈ 15,000 m^3/h at 64 °C
Exhaust air treatment	Acid scrubber, alkaline scrubber, biofilter (35 m^3 exhaust air/(m^3 biofilter × h))

5.4.3 Condensing Systems Versus Fresh Air–Exhaust Air Systems

Condensing systems recirculate drying air after it has been dehumidified in a condensing line, which is normally operated with cooling water leaving the wastewater treatment plant and heated up with the waste energy of the internal heat exchanging circuit. Fig. 5.21 illustrates a condensing line of two indirect heat exchangers marked in blue and with two additional heat exchangers, marked in red, for internal heat recovery from exhaust air to dehumidified air of the air circuit. The two blue condensing heat exchangers can also be replaced by a direct spraying condenser system with fillers, where the cooling water is in direct contact with the air and recirculated after having been filtered and cooled down.

Dryers with condensing systems are usually used at sites with sufficient wastewater treatment capacity. These dryers are characterized by higher thermal efficiency and lower electric energy consumption in comparison to fresh air–exhaust air systems. In addition, the exhaust air flow is limited to values usually less than 10,000 m^3/h, and in the case of special designs to less than 4000 m^3/h. Condensate is formed, being loaded with substances that cause chemical oxygen demand (COD) and biological oxygen demand over 5 days (BOD_5) as well as with ammonium (NH_4-N). Around 60–70 mass% of the

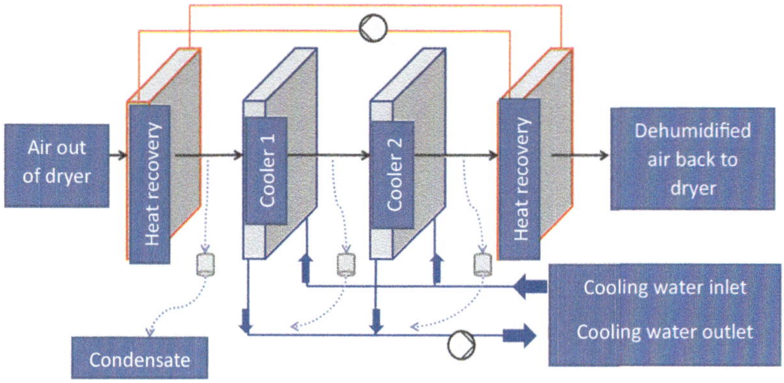

Fig. 5.21 Condensing line with indirect heat exchangers. (HUBER 2015)

ammonia (NH_3) in exhaust air is separated in the condensing system as NH_4-N. High concentrations of H_2S in dryer raw exhaust air may result in deposits of elementary sulfur at points where condensing does occur.

Direct spray condensing systems are used with fillers, operating as packed bed reactors. Fig. 5.22 shows a cross-flow packed bed condensing unit on the right side and the horizontal exhaust air pipe to the acid scrubber on the right side (HUBER 2019b).

Fig. 5.22 Cross-flow packed bed condensing unit of a two-belt sludge dryer in Austria. (HUBER 2019b)

Fig. 5.23 Direct spraying condenser without a packed bed installed on a power plant site in Germany. (Heindl 2022)

Sometimes, no packed bed is used in vertical cylindrical condensing units, as presented in Fig. 5.23, with the two taller containers in the foreground and in the background for two drying lines. The condensers are operated in counter-current flow of air and sprayed water. The columns in the center represent an acid and a caustic scrubber for each dryer.

As a rule of thumb, the heat transfer coefficient α_V related to the empty volume of the column valid for saturated gas and cooling water can be estimated as 5900–6300 kJ/(m^3 × h × K) (Schmidt 1970), and the required volume of the condenser calculated by:

$$V = \frac{Q}{\alpha_V \times \Delta\vartheta} \tag{5.8}$$

with Q being the energy for gas cooling and condensation and $\Delta\vartheta$ the average logarithmic temperature difference of gas and cooling water for counter-current processes (Schmidt 1970). The upward gas velocity is kept between 0.7 and 1.5 m/s, corresponding to a minimum water drop diameter of 0.15–0.20 mm to prevent entrainment of water drops by air. On the other hand, the water drop diameter must not exceed 0.5 mm to avoid the development of surface vibrations and the disintegration of water drops to smaller diameters (Berl 1935).

Fig. 5.24 Two-belt dryer with a fresh air–exhaust air system and integrated heat recovery in Germany. (HUBER 2021)

At sites where higher wastewater accumulation is not desired or acceptable, i.e., at industrial sites or power plants without sufficient wastewater treatment, the latter non-condensing fresh air–exhaust air system is used. The exhaust air with water vapor extracted from sludge is released to the environment after having been treated and deodorized. Thus, only the smaller amount of wastewater from the exhaust air treatment scrubbing system needs to be treated, whereas the water evaporated from sludge is removed without condensing. There is also an internal closed-loop heat exchange, in which the fresh air is heated up by exhaust energy taken from exhaust air released to the environment. Fig. 5.24 shows a belt dryer with a fresh air–exhaust air system and integrated heat recovery (HUBER 2021).

5.4.4 Summary of Belt Drying Systems

Some of the belt drying systems applied in the drying of sewage sludge use extrusion systems for the processing of dewatered sludge upstream of the dryer. The strains evenly spread on the feeding belt of the belt dryer have a fixed surface to volume ratio and ensure even drying of sludge with very low dust development. However, the sludge needs to be sufficiently dewatered to achieve a stable bed of sludge strains with sufficient through-draft of drying air.

Partial recirculation of dried sludge and mixing with dewatered sludge is also carried out almost independent of the degree of dewatering of sludge fed to the drying line, as mixed sludge has a DR content of around 60–65% to overcome the sticky phase of sludge. This mixed sludge can be dosed across the belt by a screw conveyor.

Partial recirculation of dried sludge is used in the case of sludge with a low degree of dewatering and low DR values, thus causing unstable extruded strains on the belts. In the case of partial recirculation of dried sludge, sieving of dried sludge is carried out, and sieved material is used for mixed sludge. In some cases, increasing DR to 30% makes such a sludge suitable for the extrusion process, thereby creating a mixed process. Sludge with a lower DR simply requires a higher rate of recirculation than fully dried sludge. As a rule of thumb, the rate of recirculation of dried material ranges from 65 to 70 mass% of dried sludge leaving the dryer when a DR of mixed sludge of 60–65% has to be adjusted. However, dust generation in drying needs to be considered and the cleaning process adjusted.

Whether or not condensing systems are used depends on the conditions on site, i.e., whether or not sufficient wastewater treatment capacity is available. In the case of a fresh air–exhaust air system, greater efforts in terms of exhaust air cleaning need to be accepted in favor of a lower quantity of exhaust water arising.

Belt dryers are usually supplied with waste heat—at least to a predominant extent—for reasons of environmental protection and economics.

5.5 Air Recirculation

Recirculation of drying air is carried out in all cases to save energy by increasing the water vapor load in drying air. Typically, the rate of recirculation ranges from 80 to 95% of the total mass flow of drying air. The arrangement of air heaters and fans varies from system to system. In any event, all belt dryers have a modular design with equal sections arranged in line. Most of these have a circulating fan and an air heater, supplied by hot water, steam, or thermal oil. Direct heating of recirculated air is carried out by inline direct-fired heaters. Typical arrangements of fans and air heaters in combination with airflow include:

- Air heaters arranged underneath belts and upward air flow in combination with a helical direction of air in two-belt dryers. Air is guided from one section to the following section. Thus, there is a cross co-current of air and sludge on the first belt and a cross counter-current flow for the second belt running in the opposite direction.
- Sectional circulation of air and downward airflow through sludge in two-belt dryers, where each section has its own exhaust air connection and heated air is separated into two airflows. Air heaters are usually arranged vertically beside the drying belts. Airflow is directed from a side chamber and deflected by 90 ° for vertical streaming downwards through the sludge layer on the belts.
- Airflow is directed several times from one section to the following section in a counter-current flow to the sludge in the case of single-belt dryers, where air heaters are arranged above the transport belt and airflow is directed from top to bottom through the sludge bed on the belt.

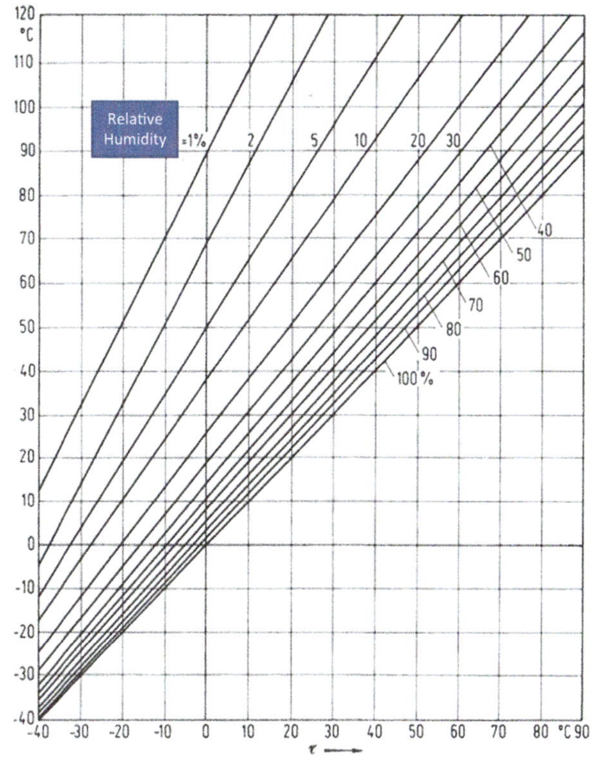

Fig. 5.25 Dew point τ of a gas/air (*horizontal axis*) depending on gas/air temperature (*vertical axis*) and relative gas/air humidity. (Hengstenberg et al. 1980)

Dryer design influences the footprint, accessibility for cleaning works, and susceptibility to dust deposition inside the dryer casing. Good accessibility is important for proper maintenance and cleaning as well as for the safe operation of belt dryers.

In any event, the rate of circulation is limited by the condensing temperature of drying air, i.e., no part of the dryer casing and internal equipment may assume a temperature below the condensing temperature (dew point). Especially in the case of outdoor installation, insulation needs to be carried out with great care, and an additional cooling effect by high wind speeds needs to be taken into account. Therefore, either the length of the dryer is restricted at a given air mass flow or airflow needs to exceed a minimum value to avoid condensation. Fig. 5.25 shows the dew point τ depending on dry air temperature and relative humidity (Hengstenberg et al. 1980).

The dew point (temperature) can be calculated by:

$$\tau = \frac{237.3 \times \left(\log \frac{x \times p}{1.33 \times (0.622 + x)} - 0.66\right)}{7.5 - \left(\log \frac{x \times p}{1.33 \times (0.622 + x)} - 0.66\right)} \tag{5.9}$$

with x as the absolute humidity of air/gas in kg/kg, p is the absolute pressure in bar and τ in °C (Mühlbauer 2009).

5.6 Insulation of Belt Dryers

In addition to dryer size and temperature of drying air, there are several factors influencing the type and thickness of belt dryer insulation:

- Necessary reduction of heat loss
- Avoidance of condensation of water vapor and additional gases inside the dryer
- Indoor versus outdoor installation, with additional heat losses and necessary rain protection.

The normal design of dryer insulation consists of 80- to 100-mm sandwich panels (metal sheet–rock wool–metal sheet), with a heat conductivity of $0.513–0.418\,\mathrm{W/(m^2 \times K)}$. Internal metal sheets are made of 304 L/1.4307 grade or 316 L/1.4404 grade stainless steel, while external metal sheets are powder-coated mild steel or grade 316 L/1.4404 stainless steel when the dryer installation is near the coast. For large belt dryers with a smaller ratio of casing surface relative to dryer volume, the heat loss is usually less than 1% of the installed thermal power. Heat loss needs to be limited for reasons of energy usage efficiency and acceptable temperature inside the dryer building.

For water evaporation capacities exceeding 3000 kg/h, dryer casings are sometimes made of steel-reinforced concrete, thereby simultaneously forming a dryer building in itself. Very large units have been installed where belt width ranged up to 8 m and dryer length up to more than 50 m. In such cases, erecting an appropriate building represents a high investment and concrete casing can be economically viable (Fig. 5.26). Of course, all piping and fans outside the concrete casing need to be adequately insulated.

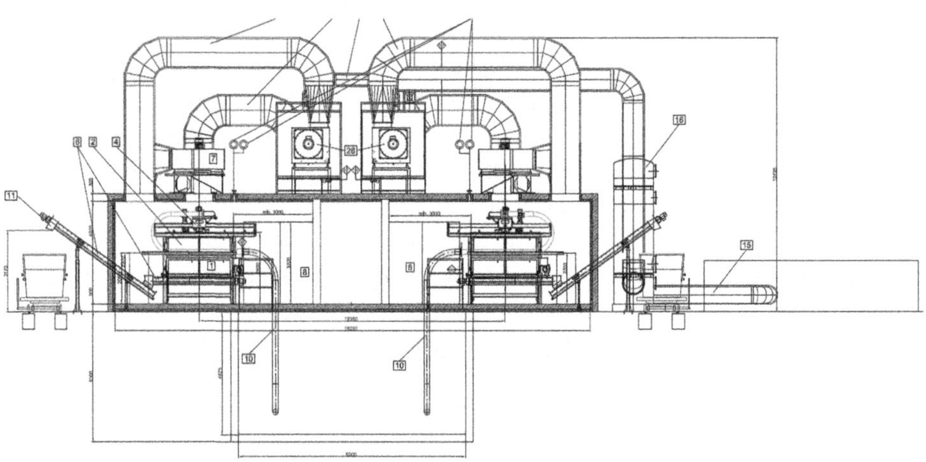

Fig. 5.26 Two dryers installed in a concrete casing with internal insulation and outdoor installation of fans and exhaust air treatment. (HUBER 2013)

Concrete itself is liable to higher process temperatures and sharp temperature drops across the thickness of the concrete wall. Both concrete tensile strength and compressive strength sharply decrease with increasing temperature. At 100 °C/200 °C, the tensile strength of concrete is 80–90%/70–85% of the values achieved at 20 °C (Walz 1962). Compressive strength starts to decrease at temperatures exceeding 200 °C. In addition to a drop in strength, crack width might exceed admissible values in the case of high temperature differences between inside and outside surfaces of the dryer concrete walls. At temperatures above 100 °C, increased coverage of reinforcement steel, thinner steel diameters, and the use of aggregates with similar thermal expansion properties to steel are recommended to avoid harmful separation of concrete and steel.

To ensure a permanent safe state of the concrete, insulation needs to be applied to the inner hot side of the concrete wall. Insulating plaster of 100 mm thickness can be sprayed on a reinforcement to achieve heat conductivity of approximately 0.06 W/(m × K) at a temperature of 150 °C. The ceiling can be covered with plates well-known from heat protection equipment for fire prevention.

The concrete foundation of the dryer is designed with foam glass insulation to prevent heat loss to the ground beneath the dryer, having a specific heat conductivity of approximately 0.05 W/(m × K).

Outdoor installations are used for very large drying units, having roofs and side panels for rain and wind protection. These designs are more liable to higher heat loss and to condensation inside the dryer; therefore, careful design and wind protection need to be ensured to avoid condensation and possible corrosion.

5.7 Energy Demand in Belt Drying of Sewage Sludge

A variety of factors influence the specific energy demand in belt drying of sewage sludge:

- Initial and final DR of sludge
- Sludge constituents that influence drying time, such as protein or fat content
- Temperature level of air
- Degree of drying air recirculation
- Air flow through the sludge bed
- Porosity of the sludge bed
- Surface to volume ratio of sludge in the bed
- Stability of the sludge bed in the case of extrusion
- Peripheral machines for sludge handling and exhaust air treatment with an effect on specific electrical energy demand.

For medium-temperature drying, i.e., sludge temperatures ranging from 50 to 100 °C and the temperature of the heat carrier at between 90 and 150 °C, typical values of specific thermal energy demand range from 800 to 900 kWh/mt of water evaporation and specific electric energy demand of between 40 and 80 kWh/mt of water evaporation. For low-temperature belt dryers with sludge temperatures below 50 °C, specific values are higher due to the higher air flow required. Specific thermal energy demands of up to 1000–1500 kWh/mt of water and a specific electric energy demand of up to 150 kWh/mt of water can be encountered for low-temperature belt drying systems.

5.8 Types of Belts

Two main types of belts are used for belt drying of sludge:

- Woven plastic belts, where the kind of plastic is adapted to the process temperature and suitable for a constant process temperature of up to 150 °C
- Conveyor chain conducted, perforated stainless steel plates, where the temperature can be adjusted up to 180 °C

The three-dimensionally woven plastic belts are made of polyethersulfone (PES) or polyphenylene sulfide (PPS) and thermally fixed on site at a temperature 40 K higher than the operating temperature. At temperatures above 100 °C, PES belts are more liable to hydrolysis and loss of strength. The temperature for continuous operation and peak temperatures as well as other design features are shown in Table 5.9. The woven structures of PES and PPS belts are described in Figs. 5.27 and 5.28. Tolerances in width are usually less than ±3 mm, and elongation of the belt in continuous operation should be less than 0.5% of the total length. The angle of the belt seam to the belt border must be 90 ° ± 0.1 °.

The plastic woven belts have the advantage of self-cleaning and are less liable to sludge sticking to the belt material. However, the running behavior of the belt needs to be well adjusted during the commissioning period and controlled during operation. The edge gradient of the belt can be monitored and side running detected. Single-belt dryers often use a special roller for automatic belt control, but the installation space requirement is high (Fig. 5.29).

Table 5.9 Plastic belts for belt dryers: technical data. (GKD 2021)

Type of plastic	Maximum operating temperature and peak temperature (short time) in °C and °F	Thickness of woven belt in mm/in	Opening (bulb diameter) in µm	Air flow through the belt at a pressure drop of 200 Pa in $m^3/(m^2 \times s)$	Wire diameter length/cross in mm and in	Weight in kg/m^2	pH-value of sludge or sludge liquid
Polyethersulfone (PES)	110/140 230/284	2.3/0.0906	490	3.4	0.65/1.00 0.0256/0.0394	1.7	<9
Polyphenylene sulfide (PPS)	150/180 302/356	2.2/0.0866	510	4.5	0.55/0.80 0.0217/0.0315	1.2	1–14

Fig. 5.27 Structure of polyethersulfone (PES) belt. (GKD 2021)

Fig. 5.28 Structure of polyphenylene sulfide (PPS) belt. (GKD 2021)

A factory acceptance test of a belt with a width of 2.6 m and more than 32 m of axis distance—carried out at the GKD works in Düren, Germany—is shown in Fig. 5.30 (Heindl 2018). A side deviation, measured at the roller in the foreground, in a belt running at that length of less than ±10 mm is good, and from ±10 to 15 mm is acceptable. Of course, these values depend on belt dryer design. Proper adjustment of the belt run during commissioning is essential for the reliable operation of a woven plastic belt. The belt run can be monitored by mechanical or optical probes.

Steel belts are made of perforated steel plates with bent side guiding plates to form a material bed. They are driven by heavy roller chains, which need constant lubrication.

Fig. 5.29 Belt tracking and guiding device for a single-belt dryer equipped with a woven plastic belt. (Heindl 2019)

Fig. 5.30 Factory acceptance test of a polyethersulfone (PES) belt with more than 60 m of running length at the GKD works in Düren, Germany. (Heindl 2018)

Since the steel plate belts tend to be liable to sludge sticking, rotating brushes are used for continuous belt cleaning. Small sludge particles might fall through the two-dimensional holes of the plates. Therefore, dust agglomerates on the dryer floor and needs to be constantly discharged or occasionally removed. The slots have a width of around 3 mm and a length of around 15–20 mm, as illustrated in Fig. 5.31 (Heindl 2000). With proper maintenance, steel plate conveyors are characterized by reliable long-term operation.

Fig. 5.31 Hinged perforated steel plate band with lateral chains on the upper carrying side and on the lower running side. (Heindl 2000)

5.9 Belt Dryer Control

Modern belt dryers are equipped with probes that measure air temperatures and pressure in the dryer to control the inside temperature and low vacuum operation. The DR content of sludge is increased from 18–35% to more than 90% for full drying. Dryer control also includes measuring the moisture of sludge fed to the dryer and leaving the dryer. Changes in the DR content of dewatered sludge at the dryer entrance are compensated by reducing or increasing sludge feed when the DR content gets lower or higher. Thus, the dryer can be kept on constant water evaporation and its capacity can be fully exploited. If the temperature of the heat carrier for air heating fluctuates, throughput also needs to be adapted to achieve full drying of the sludge.

The microwave moisture probes are based on microwaves with two different frequencies. The runtime of microwave-absorbing moisture is increased in the case of lower total solids or higher water contents in comparison to the runtime of microwaves with a frequency that is not absorbed by water molecules. The difference in runtime is a measure for the water content of sludge. Dewatered sludge is pumped through tubes and integrated ring probes, making constant in-line measurement possible. To ensure full filling of the probe, a minimum pressure of at least 1.5 bar(a) is necessary. Fig. 5.32 shows the principle of measurement and the arrangement of the probe in a tube (Valmet 2021).

Fig. 5.33 illustrates the microwave probe for DR measuring of dewatered sludge, fitted to the headpiece of a flexible tube leading to the extruding device of a belt dryer (HUBER 2013).

The final DR measurement of dehydrated sludge can be performed with a planar microwave probe introduced into the bed of sludge on the final belt, where a constant flow

Fig. 5.32 Microwave moisture measuring device for dewatered sludge, integrated in a steel tube. (Valmet 2021)

Fig. 5.33 Dry residue measuring device for dewatered sludge when pumped to the belt dryer. (HUBER 2013)

at a constant density of sludge must be ensured for reliable values. Another option is to install a planar microwave probe at the bottom of the screw conveyor casing to obtain an average DR content of dried sludge. Again, a continuous mass flow of dried sludge above the probe needs to be adjusted to obtain acceptable values. Control of the final DR of dried sludge can also be carried out by measuring the temperature drop between heated air beneath and above the sludge bed layer in the last dryer module before the dried sludge is discharged. The relationship between temperature drop and the DR of sludge needs to be calibrated according to the specific dryer system. The higher the DR of dried sludge, the lower the temperature drop.

Moisture control of dried sludge is important for safe storage in silos. Fully dried sludge with more than 90% DR is needed to avoid microbiologically induced heating of sludge during the storage period.

In addition to moisture control, temperature and operation under a low vacuum need to be measured inside the dryer. Each section has a temperature probe to control the heat supply of the individual finned tube heat exchangers. Valves controlling the flow of the heating agent are manipulated automatically. If temperatures exceed a certain limit value, they also cause cooling of circulated air and, in emergency cases, i.e., in the event of fire, flooding of the dryer with water to prevent further damage to the dryer. A low negative pressure inside the belt dryer of around 20–50 Pa prevents odor emission into the dryer hall. Therefore, the extracted airflow must always be slightly higher than the leakage airflow of the dryer.

5.10 Thermal Energy Supply for Belt Drying of Sludge

With regard to environmental protection and economic operation of belt dryers, energy sources other than oil or gas must be considered. This is essential against the background of climate change facing the world in the next decades of the 21st century.

Fig. 5.34 Fresnel collector for high temperature solar energy collection; the steam storage tank is shown on the *left side*. (Industrial Solar 2022)

5.10.1 Solar Radiation

Due to the wide range of possible process temperatures, exhaust energy sources from industrial processes or solar energy with energy storage are possible.

Areas with a high solar radiation density, such as the Middle East, the Southwestern United States, and countries near the equator, have high potential for the use of Fresnel collectors. Energy from concentrated solar radiation is used for heating thermal oil in a temperature range of around 200 °C. Fresnel collectors are manufactured in modules, for example, with each module having 22 m^2 of aperture mirror area covering a footprint of around 31 m^2. Radiation is focused on a central pipe running above a series of small tiltable mirrors. At night, stored energy from a tank is used to supply energy for the dryer. Fig. 5.34 shows three modules arranged in parallel lines with a total mirror surface of 1254 m^2. This unit, installed in Jordan, generates 1500 MWh/a, which is used for steam production.

Table 5.10 shows the mirror areas required to generate 500 and 100 kW peak at different locations around the world (Industrial Solar 2016).

Table 5.10 Mirror and footprint areas required for heating thermal oil and thermal power of 500 kW and 100 kW peak at different locations around the world

Location	Mirror area and footprint in m^2	Average annual efficiency in %	Energy collection in kWh/a
Columbia, South America, near the equator (500 kW peak)	880/1240	50	≈ 528,000
Saudi-Arabia, Middle East (500 kW peak)	880/1240	47	≈ 909,900
Southern Germany (100 kW peak)	176/250	35	≈ 61,600

Table 5.11 Design of a flue gas heat exchanger (cross flow counter-current) for heating drying air (according to a quotation)

Design of flue gas heat exchanger (tube construction without fins)	Data
Flue gas flow (in U-turn crosswise tubes)	3600 Nm^3/h
Temperature in	264 °C
Temperature out	167 °C
Pressure drop	700 Pa
Drying air flow (straight through tubes)	14,250 Nm^3/h
Temperature in	100 °C
Temperature out	125 °C
Pressure drop	200 Pa
Logarithmic temperature difference	53.1 K (calculated as pure counter-current)
Heat exchange	129 kW
Heat exchange surface	57.7 m^2 (including 10% reserve surface)
Heat transition coefficient	42.1 $W/(m^2 \times K)$
Tube material	1.4571/316 TI, up to 500 °C
Casing material	1.4301/304, up to 600 °C

5.10.2 Exhaust or Flue Gases

Exhaust gas from incineration processes or flue gases can also be used to heat drying air. However, the temperature of such exhaust gases must not fall below their dew point after heat exchange to avoid corrosion caused by mixtures of water and sulfuric acid condensate. For diesel engines, this temperature limit is around 170–180 °C, and for gas engines around 140–150 °C. Table 5.11 provides some information on the design of such flue gas/drying air heat exchangers. One needs to bear in mind that, as a rule, the surplus surface for heat exchange to compensate for the higher heat transfer resistance of deposits is only 10% according to the design of the suppliers. Therefore, frequent cleaning may be necessary—manually or automatically—and must to be carried out in order to keep the level of deposits and heat exchange on a constant value. Typically, the heat loss of such a flue gas heat exchanger ranges from 1 to 3% of the supplied exhaust heat, decreasing with increasing surface to volume ratio of the exchanger.

5.10.3 Overview of Possible Exhaust Energy Sources and Examples

There are various sources of exhaust energy with different temperature levels available at a wastewater treatment plant or from nearby industrial plants:

- Cooling water of CHPs: inlet heat exchanger of dryer 80–90 °C/outlet 70 °C. The outlet temperature is set according to the instructions of the supplier of the CHP to avoid failures in CHP operation. Temperature peaks need to be reduced by emergency air coolers.

- Cooling water of CHPs with the additional use of engine exhaust gas energy: Inlet heat exchanger of dryer 100–105 °C/outlet 70 °C with adapted heat exchanger (surplus surface) for increased temperature drop or with application of two temperature zones in the belt dyer, for example, 80 °C air temperature in zone 1 and 125 °C air temperature in zone 2.
- Exhaust gas of micro gas turbines with internal heat recovery: 270–300 °C,
- Exhaust gas of gas turbines: 400–500 °C.
- Exhaust gases of gas engines: 400–480 °C, lower temperature if exhaust gas driven turbochargers are used, minimum exhaust gas temperature after heat exchange approximately 140–150 °C.
- Exhaust gas of diesel engines: 350–400 °C, minimum exhaust gas temperature after heat exchange approximately 170–180 °C.
- Cooling air of clinker cooling in cement production: 200–250 °C.
- Cooling water of condensing vapor turbines: 60–65 °C.
- Exhaust steam of counter-pressure steam turbines with 3–6 bar absolute: 133–159 °C when saturated, usually surplus heating of steam leaving steam turbine reaches 40–70 K above the saturation temperature.
- Hot water of the organic rankine cycle (ORC) process of biomass heating stations for district heating supply: As a rule, inlet temperature to heat exchanger of 90 or 120 °C for pressure water nets.
- Cleaned water leaving the WWTP with the help of a heat pump to generate warm water of 55–60 °C. The temperature limit of minimum 8 °C for water leaving a WWTP must be observed.

It is important to check the availability of exhaust energy throughout the year to avoid bottlenecks in energy supply, for example, during the winter months, as is an issue with biogas plants. The carbon dioxide footprint also needs to be evaluated for different sources of exhaust energy and peripheral expenses to be clarified and included in the CO_2 evaluation, such as pipes and insulation or pumps for energy transfer systems. As a rule of thumb, the average specific heat capacity of flue gas from incineration processes can be estimated at around $0.151\ \text{kJ}/(\text{Nm}^3 \times \text{K})$ (Riesner 1967). For gases, the air ratio is set to 1.2 and for solid fuels to 1.4. This value can be used for initial technical and economic calculations.

For the use of steam as a waste energy carrier, several points need to be considered:

- Steam pressures normally range from 1.5 bar absolute to 6 bar absolute.
- Superheated steam with a temperature difference of 5–10 K to the temperature of saturated steam at the same pressure is favorable due to the fact that condensation in pipes to the heat exchanger is avoided. Adiabatic throttling of saturated steam, for example, from 6 bar absolute to 5 bar absolute results in a theoretical increase in energy transfer of around 3% due to the higher steam velocity in the apparatus (energy trans-

fer ≈ velocity$^{0.5}$), which overcompensates the loss in temperature difference of 3.2 K between saturated steam of 5 and 6 bar absolute.

- Typical steam velocities in pipes range from 15 to 20 m/s (49.2 to 65.6 ft/s), in extreme cases 25 m/s (82.0 ft/s) are reached. Higher velocities increase noise generation in valves and wear in the steam system. Steam with lower vapor pressure affords larger diameters of feeding pipes.
- Typical cross flow air velocities for steam heat exchangers range from 2.3 to 2.6 (7.2 to 8.5), in extreme cases up to 3.0 m/s (9.8 ft/s). The typical air pressure drop of heaters is around 200–250 Pa for finned oval pipes (30 × 14 × 1.5 mm) and up to 300 Pa (0.0435 psi) for normal circular pipes. Higher temperature differences between air inlet and outlet above 45–50 K might lead to a combination of a condensate heat exchanger as a preheating stage (sub-cooling of condensate) and a steam heat exchanger as a final heating stage (applied, e.g., to 7 bar absolute pressure for saturated steam and 2 bar absolute pressure for condensate; Balcke and Dürr 1997).
- Air heaters need to be integrated into the belt drying system in an inclined position to facilitate condensate flow.
- Condensate must be removed properly by sufficiently dimensioned pipes (two-phase flow) and reliable (mechanical) separators for condensate. Penetration of condensate into air heaters must be prevented as far as possible to avoid steam strokes and damage to the steam system.
- Re-evaporation heat of condensate must be used for preheating of drying air. Heat recovery by energy transfer from condensate to fresh air or dehumidified recirculated air improves thermal efficiency.
- A heat-transfer station with energy transfer from steam to hot water can be used, since the hot water energy supply of the dryer generates lower capital expenditure (CAPEX). However, for the same dryer size, a temperature loss of at least 5 K and as a result, a loss in water evaporation, needs to be accepted.

Fig. 5.35 presents specific features of (saturated) steam, which can be calculated according to the formulas (5.6–5.10).

Specific heat of evaporation of water vapor (steam) in kJ/kg is calculated by:

$$h_{\text{eva}} = 0.018 \times {p_s}^4 - 0.632 \times {p_s}^3 + 8.836 \times {p_s}^2 - 73.194 \times p_s + 2319.3 \tag{5.10}$$

with $R^2 = 0.9995$ and p_s pressure of saturated steam in bar absolute.

The temperature of saturated steam in °C can be described by:

$$\vartheta_s = 0.0714 \times {p_s}^3 - 1.9653 \times {p_s}^2 + 22.236 \times p_s + 81.747 \tag{5.11}$$

with $R^2 = 0.999$.

The temperature difference $\Delta\vartheta$ to 100 °C in K can be expressed with:

$$\Delta\vartheta = 0.0714 \times {p_s}^3 - 1.9653 \times {p_s}^2 + 22.236 \times p_s + 81.747 \tag{5.12}$$

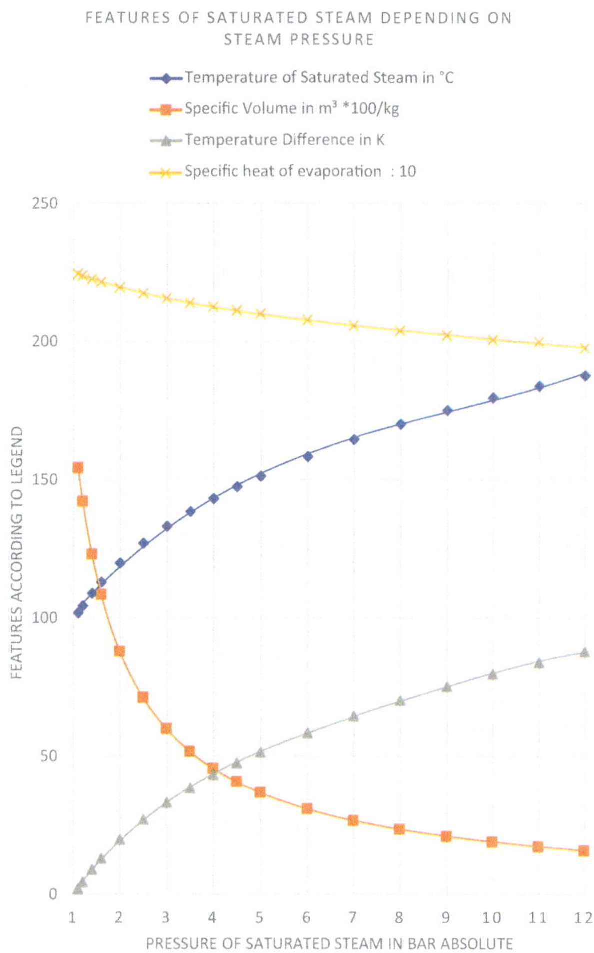

Fig. 5.35 Specific features of saturated steam. (According to data from Schmidt 1989)

with $R^2 = 0.999$.

The specific volume v of vapor in m^3/kg is defined with:

$$v = 1.6993 \times p_s - 0.00941 \tag{5.13}$$

with $R^2 = 1$.

The specific volume of vapor increases considerably with decreasing absolute pressure of saturated steam, influencing pipe and valve diameter selection. The specific steam mass flow through a valve in kg steam/($cm^2 \times h$) can be calculated by:

$$m_{steam} = 5.4291 \times p_s + 1.3897 \tag{5.14}$$

with $R^2 = 0.9998$.

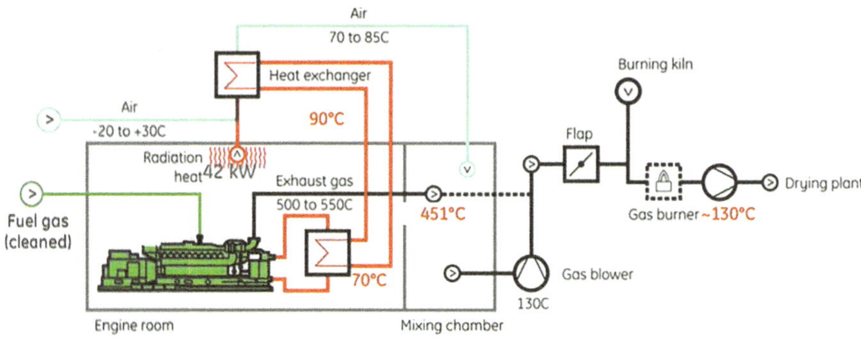

Fig. 5.36 Exhaust energy from a combined heat and power system (CHP) operated with biogas. (Elsenbruch 2010)

Fig. 5.36 illustrates exhaust energy use of gas engines for a drying process (Elsenbruch 2010). Drying air is heated up to 130 °C and exhaust gas can be optionally mixed to heated air. Fuel gas can be biogas or a mixture of biogas and natural gas. Radiation heat of gas engines is also used. In this case, the electric energy production is set to 526 kW and the recovery of thermal energy to 653 kW. For 1300 kW_{th} fed to a gas engine by fuel gas, the electric energy efficiency is 40.5% and the thermal energy efficiency is up to 50.2% (Elsenbruch 2010).

Larger belt drying systems can be supplied with waste energy from natural gas turbines, producing the total power demand of the drying system and also delivering sufficient thermal energy for sludge drying. Table 5.12 describes how a gas turbine can be integrated into

Table 5.12 Power and thermal energy supply for sludge drying using a gas turbine

Conditions	Units	Remarks
Total gas fuel supply for gas turbine	14.43 MW_{th}	
Ambient conditions for gas turbine	Air: 23 °C, relative humidity 77%, altitude of site: ≈ 1400 m above sea level	
Electric energy output of turbine	4327 kW	
Exhaust gas flow of turbine	62,916 kg/h at 524 °C	
Thermal energy for dryer as hot water 140 (7.0 bar(a))/125 °C	Maximum 10.32 MW_{th}	Maximum sludge amount per day: 438 t of dewatered sludge
Electric energy for dryer lines	Maximum 578 kW_e	Additional electric energy demand of wastewater treatment plant 3744 kW
Ratio of reuse of thermal energy for drying	71.5%	

Table 5.13 Energy situation for a 50,000 population (P) wastewater treatment plant (WWTP) with belt drying of sewage sludge

Feature	Technical data
50,000 P WWTP with digestion	Dewatered sewage sludge with 27% dry residue (DR) and 51.9 kg original substance/(P × a)
Hours of operation	8000 h/a
Two-belt dryer with condensing line	20 m^2
Drying temperature	80 °C
Water evaporation	227 kg/h
Two-gas engine With focus on thermal energy generation	Thermal efficiency: 50% Electric efficiency: 36% Thermal power: ≈ 610 kW Electric power generation: 219.6 kW Exhaust energy water 90/70 °C
Electric current consumption of a WWTP with 50,000 P (in Germany)	≈ 33 kWh/(P × a)
Electric energy consumption of the belt dryer	11.35 kWh/h or 1.82 kWh/(E × a)
Total electric energy consumption with sludge drying	≈ 35 kWh/(E × a)
Electric current generation from biogas/natural gas	≈ 17 kWh_e/(P × a) ≈ 18 kWh_e/(P × a)
Thermal energy from biogas	≈ 12 kWh_{th}/(P × a)
Thermal energy for drying	≈ 36.3 kWh_{th}/(P × a)
Thermal energy for WWTP	≈ 12.4 kWh_{th}/(P × a)
Total additional thermal energy	≈ 36.7 kWh_{th}/(P × a) → Natural gas use = 7.4 m^3/(P × a) or additional biogas from co-fermentation of 11.6 m^3/(E × a)
Annual CO_2 saving	≈ 5115 t CO_2/a

the supply of power and thermal energy for belt drying of sludge. The design of the gas turbine needs to cover the electric energy demand of the WWTP and the belt drying lines.

For medium-sized WWTPs, a combination of biogas and natural gas may be feasible to obtain enough thermal energy for sludge drying from a CHP. Table 5.13 focuses on the energy situation and production of electric energy by using a piston gas engine suitable for two-gas operation.

One must check whether co-substrate, for example, from the food industry, is available to fill the energy gap by increasing biogas production of the digester. Otherwise, natural gas or hydrogen gas needs to be used.

It is also possible to combine a power station fed by wood with a belt drying system for sludge. The exhaust energy is not only used for sludge drying but also for the facility and for digester heating. Fig. 5.37 illustrates the overall arrangement of the plant built in Albstadt, Germany (*Klärschlammverwertung* Albstadt 2012), and Table 5.14 presents its technical data.

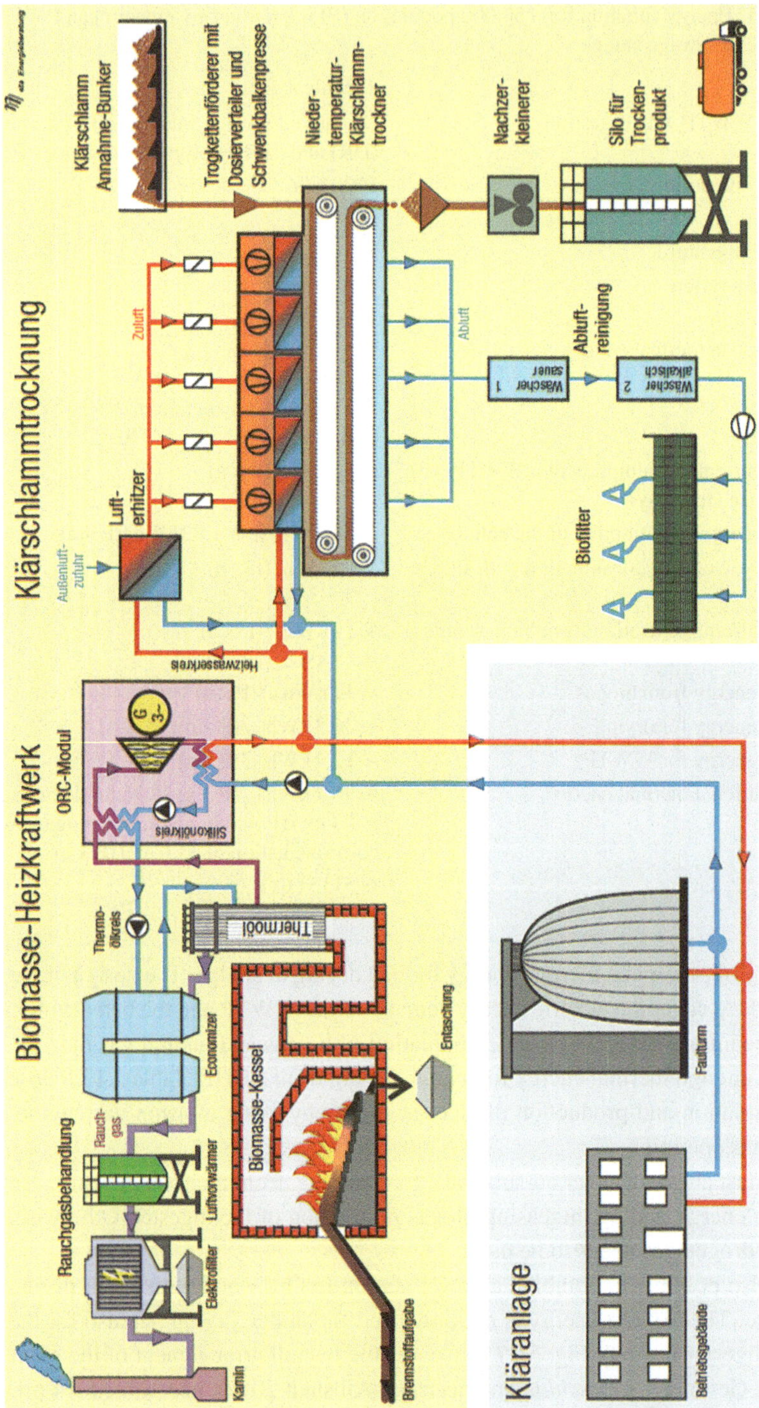
Klärschlammtrocknung
Klärschlamm Annahme-Bunker
Trogkettenförderer mit Dosierverteiler und Schwenkbalkenpresse
Nieder-temperatur-Klärschlamm-trockner
Nachzer-kleinerer
Silo für Trocken-produkt
Luft-erhitzer
Außenluft-zufuhr
Zuluft
Abluft
Abluft-reinigung
Wäscher 1 sauer
Wäscher 2 alkalisch
Biofilter
Heizwasserkreis
Biomasse-Heizkraftwerk
ORC-Modul
Silikonölkreis
Thermo-ölkreis
Thermoöl
Rauchgasbehandlung
Rauch-gas
Economizer
Luftvorwärmer
Elektrofilter
Biomasse-Kessel
Entaschung
Kamin
Brennstoffaufgabe
Kläranlage
Faulturm
Betriebsgebäude

Fig. 5.37 Combined heating and power station (wood incineration) with sewage sludge drying. (*Klärschlammverwertung* Albstadt 2012). *Brennstoffaufgabe*, solid fuel feeding; *Biomasse-Kessel*, biomass (wood) incineration; *Thermoöl*, thermal oil; *Economizer*, economizer; *Kamin*, chimney; *Elektrofilter*, electric precipitator; *Luftvorwärmer*, air preheater; *Silikonölkreis*, silicone oil circuit; *Thermoölkreislauf*, thermal oil circuit; *ORC-Modul*, organic rankine module; *Heizwasserkreis*, warm water circuit; *Außenlufzufuhr*, fresh air supply; *Lufterhitzer*, air heater; *Zuluft*, hot air supply; *Abluft*, exhaust air; *Niedertemperaturklräschlammtrockner*, low temperature sludge dryer; *Nachzerkleinerer*, final grinder; *Silo for Trockenprodukt*, silo for dried product; *Wäscher 1 sauer*, scrubber 1 acid; *Wäscher 2 alkalisch*, scrubber 2 caustic; *Abluftrenigung*, exhaust air treatment; *Faulturm*, digester; *Betriebsgebäude*, wastewater treatment plant facility; *Kläranlage*, wastewater treatment plant ◄

Table 5.14 Technical data of a combined heating and power station fueled by wood in combination with belt drying of sewage sludge (*Klärschlammverwertung* Albstadt, Germany, 2012). (SEVAR 2012)

Feature	Technical data
Sludge drying	
Dryer type	Two-belt dryer using exhaust energy of organic rankine cyle (ORC) module
Type of sludge	Digested sludge and well stabilized aerated sludge from surrounding wastewater treatment plants (WWTPs)
Mode of preparation	Extrusion with swiveling beam
Sludge throughput	≈ 12,000 mt of dewatered sludge, of which 4000 mt are from WWTP on site
Dry residue (DR) of dewatered sludge	≈ 30%
Drying air temperature	≈ 80–82 °C
Drying time	120–180 min
Throughput of dewatered sludge	1500 kg/h
Water evaporation	1000 kg/h
Drying area	≈ 90 m^2
Specific water evaporation	≈ 11.1 kg/(m^2 × h)
Wood incineration with thermal oil circuit and ORC turbine	
Wood demand	≈ 800 kg/h with DR content of 55%
Electric energy generation	≈ 300 kWh/h
Thermal exhaust energy (87/70 °C)	≈ 1450 kWh/h, demand of dryer around 850 kWh/h (estimated)
Exhaust air treatment	Two-stage scrubber and biofilter (100 m^2)
Exhaust air flow (wet)	≈ 20,000 Nm^3/h, ≈ 100% relative humidity

5.11 Minimum Standards for Belt Dryers for Sewage Sludge

For the safe and sustainable operation of a belt dryer for sludge, minimum standards are proposed:

- Corrosion resistance: Dryer parts in contact with sludge and drying air must be made of 304 or 316 grade stainless steel or of suitable plastic material. Heat exchangers in detail must be equipped with corrosion proof materials—copper- or zinc-plated steel are not suitable. Dust in drying air acts abrasively and removes coatings or zinc layers from heat exchanger surfaces. Aluminum is also not used in belt dryer design as dryer condensate is caustic with a pH of 9.5–11 as a rule, thereby destroying aluminum.
- Exhaust air treatment: The minimum equipment consists of an acid scrubber, an alkaline scrubber, and a biofilter or an activated carbon filter. Industrial sludge in general or municipal sludge that has been stored for a longer period may require additional equipment for exhaust air treatment (see Chap. 6).
- Vacuum operation and dryer tightness: The dryer must be operated under a low vacuum ($\approx$20–100 Pa) and designed to be sufficiently airtight with regard to dryer casing and feeding as well as the discharging point for drying material.
- Formation of condensate and deposits: Deposits of sludge dust in combination with the generation of condensate by frequent start and stop operations can lead to corrosive damage even to stainless steel and must be prevented. Condensate generation inside the dryer needs to be avoided as far as possible by sufficient air circulation and suitable dryer control. Dryer design with inclined metal walls of the dryer casing reduce dust accumulation, minimize wear, and maximize safety—not least with regard to fire incidents—for continuous nonstop operation.
- Heat exchangers must be installed in such a way as to ensure good accessibility for cleaning purposes. Fin distance needs to be at least 3 mm, with 4–5 mm being better in order to reduce the susceptibility of heaters to blocking and to facilitate cleaning of the heaters.

References

Balcke Dürr (1997) Communication in offer for steam air heaters. Ratingen

Berl E (1935) Chemical Engineering (In German), 3rd edn. Julius Springer, Berlin

DWA (Deutsche Vereinigung für Wasserwirtschaft, Abwasser und Abfall e.V.) (2021) Merkblatt DWA-M379 Drying of Sludge (in German). DWA, Hennef

Elsenbruch T (2010) GE Energy Jenbacher Gas Engines—From biogas to Electricity. CHP-use in operation. Presentation "FECC – GTZ Training for Biogas Design Institutes Beijing". 13.08.2010. https://energypedia.info/images/f/f1/From_Biogas_to_Electricity_-_CHP_use_in_Operation.pdf. Accessed 28 Aug 2022

Engl K (2010) Belt Drying with Sludge Incineration at ARA Tobl, Pustertal (In Italian Language) (Presentation)

GKD (2021) Combination of biomass incineration plant with sewage sludge drying (In German). Zek Kommunal 19:44–45

Haarslev Industries (2016) Company Publication: Belt Dryer BD 3000—Low-temperature drying of municipal and industrial sewage sludge, biomass and fermented residues. Bruchsal

Heindl A (2000) Photo of a Hinged Perforated Steel Plate Band

Heindl A (2015) Practice Book of Belt Drying (In German). Springer

Heindl A (2018) Photo of Factory Acceptance Test of a Woven Belt at GKD Works in Düren

Heindl A (2019) Photo of a Belt Tracking Device

Heindl A (2022) Photo of an extrusion system of a HUBER belt dryer feeding endless strains on first belt

Hengstenberg J, Sturm B, Winkler O (1980) Betriebsmesstechnik II, 3rd edn. Measuring, Driving and Controlling in Chemical Technology (In German), vol 2. Springer, Berlin Heidelberg New York

HUBER (2012) HUBER SE Photo of an extruder for dewatered sludge

HUBER (2013) HUBER SE—Design of a concrete building comprising two drier lines in parallel

HUBER (2015) Project picture of a moisture measuring probe for dewatered sludge

HUBER (2016a) HUBER Two Belt Dryer for Sludge with Condensing Unit. Berching

HUBER (2016b) Peripheral Equipment Around a Belt Dryer, Berching

HUBER (2019a) Practical Data on a HUBER Belt Dryer for Digested Sludge. Berching

HUBER (2019b) Photo of Cross-Flow Packed Bed Condensing Unit of a Two-Belt Sludge Dryer in Austria

HUBER (2021) Photo of a Fresh Air Exhaust Air Belt Dryer. Berching

Industrial Solar (2016) Email communication. Industrial Solar, Freiburg

Industrial Solar (2022) https://www.industrial-solar.de/en/references/solar-steam-generation-at-jti/. Accessed 25-08-2022

Klärschlammverwertung Albstadt (2012) Brochure on "Combined Heat and Power Station with sewage Sludge Drying" (In German). Klärschlammverwertung, Albstadt

Luggen E, Widmer F, Wuhrmann K (1976) The Particle Size Distribution as characteristic Parameter of Sludge (In German). Wasser Énerg Luft 68(11–12):257–260

Mühlbauer W (2009) Manual of Cereal Drying (In German). Agrimedia Clenze

Poxleitner M (2012) Determination of Influence of Process Parameters on Drying of Dewatered, Extruded Sludge in a Belt Dryer Simulator (In German). Hamburger Fern-Hochschule. University of Applied Sciences

Riesner W (1967) Using Exhaust Heat of Flue Gas (In German). VEB Deutscher Verlag für Grundstoffindustrie, Leipzig

Schilp R (1977) About Technology of Paste-like Substances (in German). Chem Ing Tech 49(5):374–380

Schmidt E (1989) Properties of Water and Steam in SI-Units, 4th edn. Springer, Berlin Heidelberg New York (Enlarged Printing)

Schmidt J (1970) Process of Gas Preparation (In German). VEB Deutscher Verlag für Grundstoffindustrie, Leipzig

Schubert R (2013) Determination of factors influencing extrusion of dewatered sludge (In German). Bachelor Work. Georg-Simon-Ohm University Nuremberg.

SEVAR (2012) Enterprise Publication on "Belt Dryer BD 3000/6 Albstadt" (In German). Karlsdorf

SEVAR (2013) Reference paper Opole. SEVAR, Karlsdorf

SEVAR (2022) Technical information on SEVAR belt dryer delivered by manufacturer SEVAR AG, Karlsdorf, Germany

STELA (2023) Pictures of belt dryers. Massing

SÜLZLE-KLEIN (2017) References Belt Dryer. Niederfischbach

SÜLZLE-KLEIN (2022) https://suelzle-klein.de/produkte/bandtrockner/. Accessed 10 Jan 2022

Valmet (2021) https://www.valmet.com/automation/analyzers-measurements/wastewater/total-solids-measurement-ts/. Accessed 03-09-2021

Veolia (2022a) https://www.veoliawatertech.com/sites/g/files/dvc3601/files/document/2019/01/BioCo%20Brochure%20EN.pdf. Accessed 6 Nov 2022

Veolia (2022b) https://www.veoliawatertech.com/en/publications/case-studies/biocon-reducing-1300-miles-biosolids-hauling-risks

Veolia Krüger (2012) Company Publication Biocon Sludge Drying Plant

Walz K (1962) Concrete behavior in the temperature range of 80 to 300 °C (In German). Beton 8:363–364

6 Exhaust Air and Condensate Treatment

It must be emphasized that sewage sludge is a sink for harmful and odorous substances. These do not simply vanish during the drying process, but remain either in the dried sludge or are transferred to the exhaust air and/or condensate produced. Due to this, greater efforts in terms of treatment need to be made either for exhaust air or for condensate.

Properly operated exhaust air treatment is essential for public acceptance of the sludge drying plant. It is important to mention that the prevailing wind direction and the frequency of atmospheric inversions also influence odor dispersion and, consequently, the choice of site for a drying plant, especially with regard to odor issues during abnormal operation or incidents.

The exhaust air treatment reduces odor emissions, but claims for „no odor" or „no annoyance at the boundary of a drying plant" are no options. There will be a measurable odor concentration in the off-gas of an exhaust air treatment and very often there are other odor sources nearby. Aside, the fact that removing the residual small amount of odor units affords unreasonable high CAPEX and OPEX for the exhaust air treatment, a reminant risk of nuisance will always be present and has to be accepted (according to Stuetz and Frechen 2001). After all, the further processing of sludge originating from municipal wastewater is done in the service of the population.

Often local authorities set their own regulations according to the existing odor and VOC-loads at the site. Sometimes methane contents are excluded from VOC-limits as this gas component cannot be separated from exhaust air with scrubbers, bio filters and activated carbon filters. However it has to be underlined that very low limits always afford high CAPEX and OPEX for staged exhaust air treatment units which might turn the sludge drying project unprofitable.

A. Heindl, *Belt Drying of Sludge*, https://doi.org/10.1007/978-3-662-66512-1_6

Thermal drying of sludge results in the transfer of volatile organic compounds (VOC) and inorganic gas components to the exhaust air, some of which are highly odorous even in low concentrations. The main factors influencing the odor units in exhaust air include:

- Degree of sludge stabilization
- Temperature of sludge during drying
- Type of belt dryer, i.e., low- or medium-temperature dryer or condensing and fresh air–exhaust air system
- Sludge composition with regard to ammonium (NH_4–N), volatile sulfur components
- pH-value of sludge
- Storage time running from dewatering to drying and time for sludge handling (mechanical stress [also alternating tension and pressure forces] on sludge and intermediate storage)
- Mechanical stress (also alternating tension and pressure forces) on sludge in combination with intermediate storage

The results of practical measurements with regard to odor units and ammonia contents are shown in Table 6.1 (HUBER 2020/22; ZKA Ingolstadt 2020).

Table 6.1 Practical values of odor units and ammonia contents in the exhaust air of belt dryers for sewage sludge

Belt drying system	Odor units/m^3 in exhaust air	Ammonia content in mg/m^3 (with temperature in °C)	Remarks
Three-belt dryer Low temperature of drying air 48–50 °C, fresh air–exhaust air system	3000–6000	≈ 30–45 (30–40)	Digested sludge, piston pumping, and intermediate storage upstream of drying, no external sludge
Two-belt dryer Medium temperature of drying air around 80–100 °C Condensing system	20,000–30,000	≈ 200 (35–45)	Digested sludge, no external sludge, no prolonged storage
Two-belt dryer Medium air temperature of 90–100 °C, fresh air–exhaust air system	35,000–70,000, in extreme cases more than 200,000 (measured in a sludge with an intermediate storage period of more than 2 months, a very high content of volatile acids)	≈ 300–400 (65–70)	External sludge with prolonged storage and mechanical stress through piston pumping
Two-belt dryer Medium air temperature of 130 °C, condensing system	Unknown	≈ 200–700 (45–55)	External sludge with short storage time

Table 6.2 Practical measurement of volatile organic compound (VOC) and odor units in belt drying of sludge

Type of belt dryer	One-belt dryer	Three-belt dryer
Sludge preparation	Partial back-mixing of dried sludge	Extrusion
Sludge type	Unknown, but with longer intermediate storage of dewatered sludge	Sludge of wastewater treatment plant with co-substrate feeding to digestion
Air temperature in °C	Medium ≈ 120–125	Low ≈ 58–60 °C
Throughput of fresh dry mass (DM) in kg/h	4800 (approximately 33% of total DM throughput)	480
DR_1/DR_2 in %	> 60/90*	30/> 90
Exhaust air flow in m^3/h at temperature in °C	≈ 67,000 at 60 °C	≈ 26,000 at 34 °C
VOC in mg/m^3	≈ 150–270	≈ 35
VOC in mg/kg DM	≈ 2100–3780	≈ 1080
Odor units in OU/m^3	≈ 26,000**	≈ 500–20,000***
Odor units in OU/kg DM	≈ 360,000	≈ 30,000–1,080,000

*Dryer system with partial recirculation of dried sludge (estimated ≈ 67%); **value seems to be comparatively low, typically higher values of up to 50,000 OU/m^3 and more are encountered (in extreme cases up to 6,000,000 OU/kg of DM throuphput have been measured); ***depending on co-substrate feeding to digestion

The quantity of VOC transferred to the exhaust air of belt drying systems covers a range of 500–1500 mg C/kg dry mass (DM) of dried sludge for immediate drying of sludge after dewatering at the wastewater treatment plant. However, 2000–7000 mg C/kg of DM could be measured for sludge imported from external sources and having been stored and mechanically stressed for a longer time period before entering the dryer. Extreme values might reach up to 15,000 mg C/kg of DM. The VOC concentrations in exhaust air depend on the drying system and typically range from less than 10 mg/m^3 for low-temperature belt dryers with high exhaust airflow to as much as 250 to 350 mg C/m^3 for medium-temperature dryers at the specific temperature level of exhaust air. The latter maximum values correspond to a specific concentration of VOC of up to 430 to 600 mg C/Nm^3 dry of exhaust air (HUBER 2020/22). Many VOCs have poor water solubility or are completely insoluble in water—with the exception of, for example, carbon acids of low molecular weight—and pass unchanged through condensing or scrubbing systems based on water as long as their structural stability is not damaged or destroyed. However, a certain smaller amount of them is physically absorbed.

Two practical measurements of VOC and odor units of raw exhaust air in low- and a medium-temperature belt drying of sludge were recorded and are presented in Table 6.2 (Dobslaw et al. 2010; measurements of the Umtec Institute from 2011).

In summary, belt drying of sludge from external sources at a site without wastewater treatment results in higher VOC transfer to exhaust air, expressed in mg C/kg of DM throughput. The odor units of the medium-temperature dryer are comparatively low, indicating that due to a longer storage period, odorous short-chain components have been already eliminated in the sludge and the development of volatile acids limited. However, sludge with a low degree of stabilization and consequently high content of volatile acids might cause high odor loads despite prolonged storage time. The low temperature drying of digested sludge of a wastewater treatment plant dried immediately after dewatering resulted in a far lower VOC transfer. The high variation in odor units relative to DM of dryer throughput might be explained with different feeding quantities of co-substrate during the measuring period.

Gas chromatography–mass spectrometry (GC-MS analytics; Headspace and SPME) of dewatered sludge showed a broad range of different gas components when dewatered sludge is heated to temperature level of drying air. More than 3000 chemical compounds have been identified; for over 100 of these, a mass portion of more than 1% could be measured in the gas phase (HUBER 2020/22).

6.1 Dust in Exhaust Gases Leaving Belt Dryers

Dust in sludge drying is defined as having a particle diameter of less than 500 µm in exhaust air. The dust concentration for belt dryers operating with the extrusion of dewatered sludge is, as a rule, less than 3 mg/Nm^3dry in raw air leaving the dryer. A mass balance for a two-belt dryer operating with sludge extrusion resulted in a dust transfer to exhaust air of around 60 mg of dust per kg of DM of dryer throughput. Of course, these values are related to full drying to a dry residue (DR) content of less than 94% and to a specific drying system. Also, dryer design and airflow speed through the bed of sludge influence dust development, and drying to higher DR contents will increase dust content in exhaust air. As a rule, dryers with back-mixing of a portion of dried sludge using sieved fines reach higher values of dust in raw exhaust air and are commonly equipped with dust-measuring devices in air tubes for safety reasons.

Dust from the drying of municipal sludge is hydrophilic in most cases. Therefore, its separation in condensers and/or scrubbers operating with water is effective. Dust made up mainly of black coal or lipophilic substances is not adequately separated. As a rule of thumb, dust particles with a diameter of less than 0.4–0.5 µm are taken with exhaust air along the streamlines without coming into contact with water droplets in the scrubbers. They pass through the wet cleaning systems (Knop et al. 1972).

6.2 Factors Influencing Gas Components in Exhaust Air

As mentioned above, sludge characteristics and sludge handling have a major impact on the emission of VOCs into drying air. Therefore, the cost of exhaust air treatment depends largely on sludge composition, maximum temperature of the sludge during the drying process, on the degree of sludge stabilization, on the pH-value of dewatered sludge, on mechanical stress exerted on the sludge, and on oxygen supply to sludge as well as on the intermediate storage period until the drying process is started.

Exhaust air treatment for belt drying systems is always necessary and its basic equipment for the drying of in-house-produced, well-stabilized dewatered sludge with a pH-value of 7.2 to above 8 consists of an acid scrubber, a caustic scrubber, and a biofilter. Lower degrees of stabilization, high shear forces on sludge cells during sludge conveying, intermediate storage after the application of force, and receiving sludge from other wastewater treatment plants with longer storage and transport periods will increase the quantity of VOCs in general and of odor-bearing compounds in the exhaust air. Consequently, more efforts need to be made to reduce VOC and odor loads in exhaust air to meet local requirements.

At the beginning of this century, the opinion was widespread that low temperature drying, i.e., at air temperatures of less than 55 °C, and high air flow through the bed of sludge would limit odor components in exhaust air. It was assumed—and confirmed in some early projects—that drying of freshly dewatered sludge with low air temperatures did not require exhaust air treatment. However, with growing experience in several projects, it turned out that with increasing shear forces on sludge cells, exerted by mechanical dewatering in centrifuges instead of belt presses and by piston pumps—with high pressures of up to and more than 50 bar or reverse tensioning and pressing force exerted on sludge—odor issues occurred. This was related to cell rupture, the formation of organic acids by hydrolysis of cell ingredients (e.g., propionic and butyric acid), and to ongoing microbial activity during intermediate storage before drying. The effect of cell rupture during pumping is reinforced by the increased DR of dewatered sludge due to a higher level of friction between sludge particles themselves and between the particles and the tube walls. Even for a well-stabilized digested sludge, an increase in organic acids, expressed as acetic acid equivalent, by a factor of 7–11 could be detected when comparing sludge immediately after dewatering with sludge piston-pumped and stored in a bunker before drying in samples re-suspended to a total suspended solids (TTS) content of 3%. Thus, the acetic acid equivalent increased from 134 mg/L after dewatering to 268 mg/L before pumped sludge was stored in a 30-m^3 bunker and to 805 mg/L after extrusion when sludge entered a low temperature belt dryer (Kopp 2006). Therefore, even these low-temperature dryers needed to be equipped with an exhaust air treatment system.

6.3 Types of Gas Components in Exhaust Air of Belt Dryers

Various components might be found in the exhaust air of a convective belt dryer for sewage sludge. The following tables show different types of inorganic and organic gases analyzed in the exhaust air of belt dryers or detected by gas chromatography (GC) in combination with mass spectroscopy (MS) on dewatered sludge samples. Fig. 6.1 shows a GC-MS instrument of this kind at the University of Stuttgart, ISWA, Germany (Dobslaw 2021).

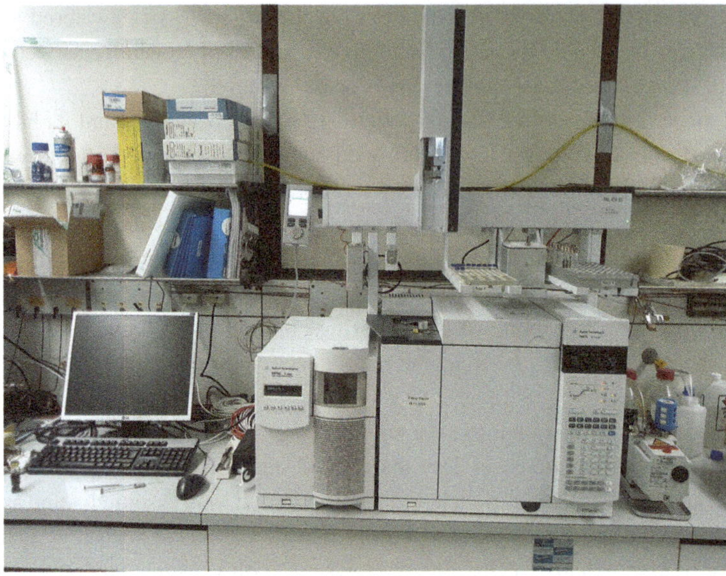

Fig. 6.1 Gas chromatography-mass spectroscopy instrument to detect gas components emerging from dewatered sludge samples when heated, installed at the ISWA, University of Stuttgart, Germany. (Dobslaw 2021)

Of course, the appearance and concentration of gas components in dryer exhaust air depend on several factors, including:

- pH-value of dewatered sludge
- Ammonia content in dewatered sludge
- Highest temperature of sludge and intensity of water evaporation during the drying process
- Content of organic sulfur in sludge (roughly determined by deducting sulfate content from total sulfur content)
- General content of volatile total organic carbon (TOC = VOC) in sludge according to TOCs in wastewater intake that are not or only partially biologically degraded
- History of sludge handling with regard to storage, mechanical conveying, and mechanical stress in combination with oxygen intake into the sludge mass.

For some gas components, ranges for odor thresholds are indicated, as odor impressions are subjective. Odor thresholds indicated in a concentration in mg/m^3 for all presented gas compounds are taken from Schön and Hübner (1996), Ruth (1986), ATV (1994), (https://pubchem.ncbi.nlm.nih.gov), (https://gestis.dguv.de), and Odor Measurement Review (Nagata n.d.). Values marked with (S) are standardized human olfactory thresholds according to Devos, calculated from literature sources with special weighing coefficients (Devos et al. 1990). These values have higher reliability.

Table 6.3 Volume (V) per mole for an ideal gas depending on temperature

Temperature in °C	V in L/mole
0	22.41
10	23.23
15	23.64
20	24.05
25	24.46
30	24.87
50	26.51
70	28.15
100	30.61
125	32.67
150	34.72

The odor index of a gas component is defined as (Täffner 2011):

$$C_{\text{index}} = c_{\text{gas}}/O_{\text{threshold}} \tag{6.1}$$

and indicates the portion of odor impression of an odorous gas mixture. The smaller the odor threshold $O_{\text{thereshold}}$, the higher the odor impression of a gas component at an equal gas concentration.

The gas concentration c in g/m^3 can be calculated from values given in ppm v/v using the formula:

$$c_{(\text{g/m}^3)} = (c_{(\text{ppm})}/10{,}000) \times 10 \times M/V \tag{6.2}$$

with M as molar mass in g/mole and V as volume in L per one mole of gas at a given temperature.

Table 6.3 shows values of V for different temperatures, calculated for an ideal gas.

The transfer of a gas concentration c under operational conditions, i.e., at temperature T in K and overall pressure p in Pa to a normalized wet or dry gas concentration c_{wet} and c_{dry} on the basis of a normalized temperature $T_{\text{N}} = 273.15\,K$ (0 °C) and an overall normalized pressure $p_{\text{N}} = 101{,}325$ Pa is described by the formulas:

$$c_{\text{N wet}} = c \times \left(\frac{p_{\text{N}}}{p}\right) \times \left(T_{\text{oper}}/T_{\text{N}}\right) \tag{6.3}$$

$$c_{\text{N dry}} = c \times \left(\frac{p_{\text{N}}}{p} - p_{\text{D}}\right) \times \left(T_{\text{oper}}/T_{\text{N}}\right) \tag{6.4}$$

$$p_{\text{D}} = x \times \frac{p}{0.622 + x} \tag{6.5}$$

where p_{D} is water vapor pressure at the temperature of operation T_{oper}, all temperatures are meant in K, and x is the absolute humidity of gas in kg water vapor/kg of dry gas. In

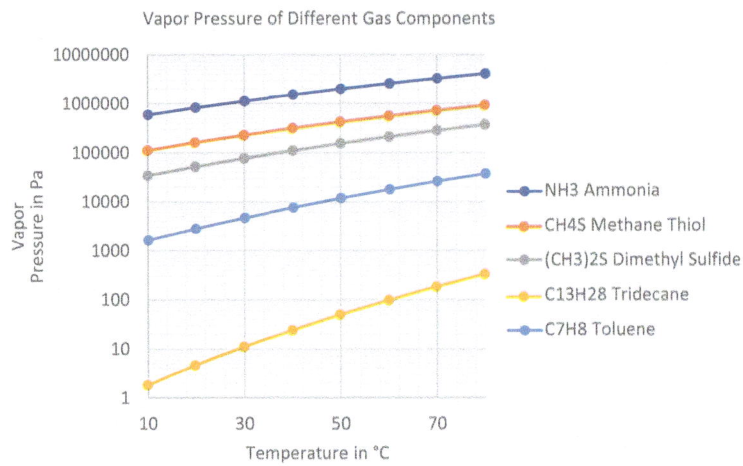

Fig. 6.2 Vapor pressure of different gas components in exhaust air depending on temperature

the German *TA Luft* (Technical Instructions for Keeping Air Clean), the odor values are normalized to wet gas at 20 °C (293.15 K or 68 °F) and 101,325 Pa.

The higher the vapor pressure of a gas component, the higher its volatility. The vapor pressure in mm Hg can be calculated by using the Antoine equation, depending on temperature ϑ in °C:

$$\log_{10} p_{\mathrm{V}} = \mathrm{A} - \mathrm{B}/(\vartheta + \mathrm{C}) \quad \text{or} \quad p_{\mathrm{V}} = 10^{[\mathrm{A}-\mathrm{B}/(\vartheta+\mathrm{C})]} \tag{6.6}$$

Fig. 6.2 illustrates the temperature dependence of vapor pressure of different gas components in exhaust air. The parameters of A, B, and C are indicated in the following tables describing gas components occurring in the exhaust air of dryers and their features (Yaws 2015). The transfer from mm Hg to Pa is carried out by multiplying the values with 133.322. Despite the fact that tridecane has a far lower vapor pressure in comparison to toluene (consider log-scale of vertical pressure axis), it can be detected in a comparable mass concentration in GC-MS analysis of sludge samples performed at usual drying air temperatures.

The theoretical maximum volatility F of a gas component in air out of a wet sludge surface, expressed in mg/m^3, is calculated by (Summer 1970):

$$F = \frac{273.15 \times p \times M \times 10^6}{22.41 \times (273.15 + \vartheta) \times 101{,}325} \tag{6.7}$$

where p is vapor pressure of the gas component in Pa at a given temperature, M is molar mass of the gas component in g/mole, and ϑ is temperature in °C.

Calculating the theoretical volatility for ammonia (NH_3; molar mass 17.03 g/mole), methane ethiol (CH_4S; molar mass 48.11 g/mole), dimethyl sulfide ($(CH_3)_2S$; molar mass

Table 6.4 Theoretical calculation of maximum volatility for different gas components occurring in exhaust gas of sludge drying for a sludge temperature of 60 °C

Gas component (boiling point)	NH_3 (−33 °C)	CH_4S (6 °C)	$(CH_3)_2S$ (37 °C)	$C_{13}H_{28}$ (234 °C)
Vapor pressure at 60 °C in Pa	2,690,838	595,016	109,191	108,4
Maximum theoretical volatility in mg/m^3 at 60 °C	16,546,461	10,336,343	2,449,971	7,216

62.14 g/mole), and for tridecane ($C_{13}H_{28}$; molar mass 184.36 g/mole) for a temperature of 60 °C yields the results presented in Table 6.4. The situation in belt drying of sludge may result in a maximum wet sludge temperature of around 60 °C.

Table 6.4 clearly expresses that volatility rises with vapor pressure at a given temperature and decreases with increasing molar mass and boiling point temperature. As a rule of thumb, volatility rises in the temperature range from 30 to 230 °C by 50–60% for a temperature increase of 10 K (Summer 1970). The higher the boiling point temperature, the lower the corresponding volatility of a liquid chemical substance. As long as the surface of sludge granules or the sludge bed are wet, the evaporation of volatile substances is considered to be similar to the evaporation of a liquid surface. Calculated values are maximum values assuming that wet surfaces can deliver the substances without limitation, which in the practice of sludge drying does not happen.

Gas components were selected according to the GC-MS analysis of numerous dewatered sludges at temperatures equal to the applied air temperatures (ranging from 74 to 125 °C; HUBER 2020/22).

The boiling temperature of most VOC gas components is higher than the boiling temperature of water at ambient pressure. The idea that components with higher boiling temperatures are physically carried away by the water vapor during drying is incorrect. Considering that each component in the sludge has a certain vapor pressure corresponding to the sludge temperature, it is clear that all components will be transferred to the gas phase at a certain concentration, as allowed by the balance between the liquid phase in the wet sludge and the gas phase in the drying air of this component. In belt drying of sewage sludge, the maximum air temperature is limited to 125 °C (257 °F) in most cases for safety reasons. During the first constant rate period of drying, characterized by a wet surface of sludge particles, the main transfer of gas components to drying air occurs and the sludge temperature reaches approximately wet-bulb temperature of air. Assuming a maximum absolute water content of drying air of 250 g/kg of dry air (1750 gr/lbs), the wet-bulb temperature will not exceed 71 °C (159.8 °F). For extruded sludge, the end of the first drying phase depends on air temperature and air humidity, decreasing with less aggressive drying conditions, i.e., higher absolute humidity and lower dry air temperature. Values between 24 and 33% DR (67–76% of water content) for the end point of the first constant rate period are comparatively common (Tejada Gutierrez 2011). There is no distinct boundary between the phase of constant drying rate and the phase of decreasing drying

Table 6.5 Inorganic gases in exhaust air of convective dryers

Inorganic gases Chemical formula, molar mass, and boiling temperature Value of Antoine equation	Odor impression	Odor threshold in mg/m^3 at 20 °C/68 °F	Water solubility in mg/L Log K_{OW}
Ammonia (NH_3) 17.03 g/mole −33 °C A = 7.58743, B = 1013.7185, C = 248.83	Pig-like	4.0738 (S)	531,000 (20 °C) 0.23
Trimethylamine (C_3H_9N) 59.11 g/mole 2.9 °C A = 7.00312, B = 990.9392, C = 238.535	Fishy	0.0059 (S)	Unlimited 0.16
Indole (C_8H_7N) 117.15 g/mole 254 °C A = 7.13700, B = 1841.3421, C = 178.577	Fecal	0.000155 (S)	2800 (20 °C) 2.14
Skatole (C_9H_9N) (3-methylindole) 131.17 g/mole 265 °C A = 7.27498, B = 1967.1204, C = 182.116	Fecal, in low concentration perfume impression	0.00309 (S)	450 (20 °C) 2.6
Hydrogen sulfide (H_2S) 34.1 g/mole −60 °C A = 7.11958, B = 802.2266, C = 249.61	Spoiled eggs	0.0257 (S)	3980 (20 °C) Not applicable

rate, and the described effects will even range to lower DR of the sludge during the drying process. Thus, for two-belt dryers, where the DR at the turnover point from the first to the second belt is as a rule between 45 and 55% DR (Heindl 2015), it can be assumed that the effects are valid approximately for the whole sludge mass on the first belt.

Water solubility is indicated for the evaluation of the use of water-based chemical scrubbers. One must bear in mind that increasing scrubber temperature results in decreasing water solubility of the gas compound. Low log K_{OW} values (octanol–water distribution coefficient of gases) indicate higher polarity and water solubility, while high values result in higher solubility in fatty substances and very low or no water solubility. Log K_{OW} values less than ≈ 1.0 show acceptable water solubility of the gas component with regard to the use of chemical scrubbers, and values higher than 2.0 are practically insoluble in water.

The following tables divide gas components in dryer exhaust air into different categories: inorganic gases (see Table 6.5) and VOCs. The latter are divided into organic substances on a sulfur basis, organic acids, alcohols, hydrocarbons and aromatic compounds, aldehydes, and ketones (see Tables 6.6–6.11). Most of the data are taken from https://pubchem.ncbi.nlm.nih.gov.

Table 6.6 Sulfuric VOC

Organic gases containing sulfur (S) Chemical formula, molar mass, and boiling temperature Values of Antoine equation	Odor impression	Odor threshold at 20 °C/68 °F (ambient temperature) in mg/m^3	Water solubility in mg/L Log K_{OW}
Methane thiol (methyl mercaptan) (CH_4S) 48.11 g/mole 6 °C A = 7.07302, B = 1008.9751, C = 234.729	Spoiled vegetable, sulfide-like	0.00209 (S)	15,400–23,300 (25–20 °C) 0.78
Ethanethiol (ethyl mercaptan) (C_2H_6S) 62.13 g/mole 35 °C A = 7.12083, B = 1193.8233, C = 245.711	Leek-like, pungent, foul smelling	0.00282 (S)	15,600 (25 °C), strongly depending on pH-value 1.27–1.50
Dimethyl sulfide [$(CH_3)_2S$] 62.13 g/mole 37 °C (different chemical structure in comparison to ethyl mercaptan) A = 7.07585, B = 1150.8686, C = 236.491	Spoiled cabbage	0.00589 (S)	2000 (20 °C) 1.05
1-Butanthiol ($C_4H_{10}S$) 90.19 g/mole 98 °C A = 7.58185, B = 1535.0872, C = 228.692	Skunk-like, spoiled onions	0.004265 (S)	597 (20 °C) 2.28
2-Methylpropane-2-thiole (tert-butyl-mercaptan) [$(CH_3)_3CSH$] 90.19 g/mole 64 °C A = 7.5600, B = 1413.3999, C = 238.211	Camphor like	0.00126 (S)	4200 (25 °C) 2.14
Dimethyl disulfide [$(CH_3)_2S_2$] 94.20 g/mole 110 °C A = 7.25695, B = 1502.7256, C = 233.641	Stinkhorn, garlic	0.05	2500 (20 °C) 1.77
Dimethyl trisulfide [$(CH_3)_2S_3$] 126.26 g/mole 170 °C	Cooked cabbage, foul smelling	0.062–0.01	Insoluble 1.94

Table 6.7 Volatile (organic) acids

Organic acids (carbon/fatty acids) Chemical formula, molar mass, and boiling temperature Values of Antoine equation	Odor impression	Odor threshold in mg/m^3 at 20 °C (68 °F)	Water solubility in mg/L Log K_{OW}
Acetic acid ($C_2H_4O_2$) 60.052 g/mole 118 °C A = 7.27594, B = 1327.1634, C = 183.913	Acetic	0.14454 (S)	Unlimited −0.17

Table 6.7 (Continued)

Organic acids (carbon/fatty acids) Chemical formula, molar mass, and boiling temperature Values of Antoine equation	Odor impression	Odor threshold in mg/m³ at 20 °C (68 °F)	Water solubility in mg/L Log K_{OW}
Propanonic acid ($C_3H_6O_2$) 74.080 g/mole 141.2 °C A = 7.68217, B = 1730.2269, C = 219.512	Bad breath	0.03548 (S)	370,000 (20 °C) 0.33
Butyric acid ($C_4H_8O_2$) 88.11 g/mole 163.5 °C A = 7.71423, B = 1786.2007, C = 206.702	Spoiled butter, sour	0.001–0.014	Unlimited 0.79
Valeric acid ($C_5H_{10}O_2$) 102.130 g/mole 186 °C A = 7.64852, B = 1842.0816, C = 200.516	Sweaty, cheesy	0.026	25,000 (25 °C) 1.39
Caproic acid ($C_6H_{12}O_2$) 116.158 g/mole 205 °C A = 7.55390, B = 1769.4739, C = 174.802	Goaty, rancid, cheesy	0.01259 (S)	9700 1.92

Table 6.8 Alcohols

Alcohols as organic gases Chemical formula, molar mass, and boiling temperature Values of Antoine equation	Odor impression	Odor threshold at 20 °C (68 °F) in mg/m³	Water solubility in mg/L Log K_{OW}
Methanol (CH_4O) 32.04 g/mole 64.7 °C A = 8.08404, B = 1580.4585, C = 239.06	Alcoholic, pungent	186.208 (S)	Unlimited 0.74
1-Propanol (C_3H_8O) 60.01 g/mole 97 °C A = 7.82517, B = 1482.1331, C = 217.413	Alcoholic, pungent	6.0256 (S)	Unlimited 0.20
1-Butanol ($C_4H_{10}O$) 74.12 g/mole 117.7 °C A = 7.30130, B = 1285.0227, C = 173.247	Wine-like, banana-like, slightly rancid	1.51356 (S)	77,000 (20 °C) 0.84
3-Methyl-1-butanol ($C_5H_{12}O$) 88.15 g/mole 131 °C A = 7.21542, B = 1333.4601, C = 169.781	Sour, pungent	0.16218 (S)	28,000 (20 °C) 1.16–1.28
1-Decanol ($C_{10}H_{22}O$) 158.28 g/mole 230 °C A = 7.94534, B = 2145.9319, C = 191.869	Sweet, fatty, waxy, orange	0.01820 (S)	37–40 (at 25–20 °C) 3.07
1-Dodecanol ($C_{12}H_{26}O$) 186.34 g/mole 259 °C A = 8.09507, B = 2313.9698, C = 182.728	Amine-like, floral	0.12023 (S)	4 (25 °C) 5.13

Table 6.9 Hydrocarbons

Hydrocarbons as organic gases Chemical formula, molar mass, and boiling temperature Values of Antoine equation	Odor impression	Odor threshold in mg/m^3 at 20 °C (ambient temperature)	Water solubility in mg/L Log K_{OW}
Methane (CH_4) 16.04 g/mole –161.6 °C A = 6.84377, B = 435.4534, C = 271.361	No odor	No odor	22 (25 °C) 1.09
Decane ($C_{10}H_{22}$) 142.29 g/mole 174.1 °C A = 7.22035, B = 1694.6279, C = 216.459	Petrol-like	4.3652 (S)	Practically insoluble 5.01
Dodecane ($C_{12}H_{26}$) 170.34 g/mole 216.2 °C A = 7.24104, B = 1810.4995, C = 199.381	Petrol-like	14.4544 (S)	Practically insoluble 6.10
Tridecane ($C_{13}H_{28}$) 184.37 g/mole 234 °C A = 7.34299, B = 1958.0087, C = 203.429	Petrol-like	16.5959 (S)	Practically insoluble 7.56
Tetradecane ($C_{14}H_{30}$) 198.39 g/mole 253.6 °C A = 7.38975, B = 1954.0312, C = 183.519	Mildly waxy	Unknown	2.2–3.3 × 10^{-4} mg/L (25 °C) practically insoluble 7.20
Pentadecane ($C_{15}H_{32}$) 212.42 g/mole 270.7 °C A = 7.42974, B = 2026.6694, C = 178.677	Slightly like crude oil	Unknown	Practically insoluble 7.71
Undecane ($C_{11}H_{24}$) 156.31 g/mole 196 °C A = 7.18058, B = 1706.2645, C = 200.897	Slightly petrol-like	7.76247 (S)	Practically insoluble 6.54
Styrene (C_8H_8) 104.15 g/mole 145 °C A = 6.92098, B = 1437.8521, C = 210.039	Sweet-ish, burnt plastic	0.1516	240 (20 °C) 3.05
Toluene (C_7H_8) 92.14 g/mole 110.6 °C A = 7.13657, B = 1457.2871, C = 231.827	Glue, solvent-like	5.88844 (S)	520 (20 °C) 2.73
p-Xylene (C_8H_{10}) 106.17 g/mole 138.3 °C A = 7.15471, B = 1553.9461, C = 225.230	Plastic- and solvent-like	2.13796 (S)	160 (25 °C) 3.15

Table 6.9 (Continued)

Hydrocarbons as organic gases Chemical formula, molar mass, and boiling temperature Values of Antoine equation	Odor impression	Odor threshold in mg/m^3 at 20 °C (ambient temperature)	Water solubility in mg/L Log K_{OW}
Naphthalene ($C_{10}H_8$) 128.17 g/mole 218 °C A = 6.92852, B = 1561.3175, C = 168.879	Mothball-like	0.07943 (S)	31 (20 °C) 3.30
D-Limonene ($C_{10}H_{16}$) 136.24 g/mole 177.6 °C A = 7.06744, B = 1691.1486, C = 227.441	Citrus-like	2.45471 (S)	14 (25 °C) 4.57
Phenol 94.11 g/mole (C_6H_6O) (aromatic) 181.7 °C A = 7.36683, B = 1629.4035, C = 181.369	Typically medicinal/sweet	0.42658 (S)	84 (20 °C) 1.46
p-Cresol (C_7H_8O) 108.14 g/mole (aromatic) 202 °C A = 7.57676, B = 1886.6598, C = 199.914	Horse stable-like	0.00832 (S)	20 (20 °C) 1.98

Table 6.10 Aldehydes

Aldehydes as organic gases Chemical formula, molar mass and boiling temperature Values of Antoine equation	Odor impression	Odor threshold at 20 °C (68 °F) in mg/m^3	Water solubility in mg/L (25 °C) Log K_{OW}
Heptaldehyde (1-heptanal) ($C_7H_{14}O$) 114.19 g/mole 152.8 °C A = 7.66164, B = 1756.6469, C = 213.586	Fruity, oily, fatty	0.02291 (S)	1250 2.29
Buthyraldehyde (butanal) (C_4H_8O) 72.11 g/mole 75 °C A = 7.21500, B = 1317.4444, C = 229.15	Pungent-smelling	0.02754 (S)	71,000 (20 °C) 0.88
Propionaldehyde (propanal) (C_3H_6O) 58.08 g/mole 49 °C A = 7.26342, B = 1277.1176, C = 242.556	Pungent-smelling	0.06546 (S)	679,000 0.59
Acetaldehyde (ethanal) (C_2H_4O) 44.05 g/mole 20 °C A = 7.31939, B = 1176.3175, C = 244.272	Fruity, orange-like, pungent	0.338844 (S)	1,000,000 −0.34
Benzaldehyde (C_7H_6O) 106.13 g/mole 179 °C A = 7.25044, B = 1733.7871, C = 217.931	Bitter almond	0.18621 (S)	6950 1.40

Table 6.11 Ketones

Ketones as organic gases Chemical formula, molar mass, and boiling temperature	Odor impression	Odor threshold at 20 °C (68 °F) in mg/m^3	Water solubility in mg/L (25 °C) Log K_{OW}
Acetone (propanone) (C_3H_6O) 58.08 g/mole 56.1 °C A = 7.31742, B = 1315.6735, C = 240.479	Slightly sweetish odor, nail polish remover	34.6737 (S)	Unlimited −0.24
Butanone (butyraldehyde) (C_4H_8O) 72.11 g/mole 79.5 °C A = 7.21500, B = 1317.444, C = 229.115	Fruity, nail polish remover	23.44229 (S)	290,000 (20 °C) 0.88
Diacetyl (vinyl acetate) ($C_4H_6O_2$) 86.09 g/mole 87.5 °C A = 7.34212, B = 1393.3201, C = 239.762	Buttery odor	0.015849 (S)	220,000 (20 °C) −1.34
Camphor ($C_{10}H_{16}O$) 152.23 g/mole 205.7 °C A = 6.90229, B = 1635.5504, C = 197.554	Woody, eucalyptic	0.32359 (S)	1600 2.38

Organic compounds in dryer exhaust air always include carbon (C) and are named as volatile organic compounds (VOC). They are the volatile portion of total organic carbon contents (TOC) transferred to drying air in convective drying processes.

Organosulfuric substances can be partially separated in a caustic scrubber as long as their water solubility is sufficient. Other sulfuric compounds are removed from exhaust air in a biofilter or an activated carbon filter or need to be oxidized by UV light or regenerative thermal oxidation.

Typical values of volatile acids in the gas phase of headspace GC-MS analysis of dewatered sludge are around 2 mass%; however, values of more than 4 mass% are possible in cases of prolonged sludge storage. These data are related to total VOC. Organic acids are separated in the caustic scrubber as their water solubility is, as a rule, high enough. If not separated by a scrubber, organic acids might reduce the life of a subsequent activated carbon filter.

Silicone compounds such as siloxanes, for example, cyclotetra siloxane ($H_8O_4Si_4$) can also be found in dryer exhaust air. These components form SiO_2 and coating deposits on surfaces when oxidized. This needs to be considered in the case of the use of UV radiation lamps or a heat exchanger of regenerative thermal oxidation (RTO) units. Typically, silicone compounds can be detected in dryer exhaust air—according to headspace measurements—of around 1 mass%, expressed as Si, and maximum values of up to 2 mass% are possible.

Most of the smaller alcohol compounds show high or at least sufficient water solubility and can be removed from exhaust air in scrubbers.

Virtually all hydrocarbons are insoluble in water. They form the largest part of VOCs that cannot be separated in scrubbers in a sufficiently manner. Tridecane and tetradecane are the most common hydrocarbons in exhaust air; their share is usually up to 4 mass% relative to total VOC, expressed as C, but values of up to more than 20 mass% may be encountered in some cases.

It must be emphasized that methane (CH_4) cannot be economically separated from dryer exhaust air with scrubbers, biofilters, or activated carbon filters. Some methane can be physically adsorbed in activated carbon for a limited period of time, but it is easily displaced by larger gas molecules. Even UV-units working with ultraviolet radiation lamps of a wave length of 185/254 nm are not able to destroy more than 10% of the methane concentrated in exhaust air due to the fact that the wavelength of maximum absorption of methane is around 120 nm. Excimer UV lamps of 172 nm wavelength might be useful in further reducing the methane content of exhaust air in the near future.

Aldehydes commonly show sufficient water solubility and can be efficiently separated in water-based scrubbers.

Separation of aldehydes and ketones is sometimes carried out in a final "neutral" scrubber operating at a pH-value of around 7 and using sodium thiosulfate ($Na_2S_2O_3$) as a chemical.

The gas transfer to the drying air is enhanced by the sludge temperature and influenced by the pH-value of dewatered sludge. The nitrogen-based gas components such as ammonia or trimethylamine appear in a higher quantity when sludge is digested and the pH-value of dewatered sludge exceeds 8. In this case, a greater proportion of the sulfur-based odor-carrying components are bound as salts in the sludge and are not released to the drying air. Sludge with pH-values < 7.5 going down to 5.5, such as aerated or waste activated sludge, show a higher amount of sulfuric gas components in the dryer exhaust air, whereas the nitrogen-based substances decrease. In dryers with a high of air circulation, as in the case convective of belt dryers, hydrogen sulfide diffusing out of sludge with a lower pH-value immediately reacts with the oxygen in the drying air, forming elementary sulfur in an S8 configuration.

$$2\,H_2S + O_2 \Leftrightarrow S\Downarrow + 2\,H_2O \tag{6.8}$$

As water is also generated, this sulfur deposits principally on cooler or cooled surfaces such as those in the condensing unit of the dryer.

6.4 Regulations for Exhaust Air Limits of Gas Components in Belt Drying of Sludge

Of course, local limits for specific gas components and odor units in exhaust air released to the environment need to be observed and maintained to obtain permission for the operation of a belt drying line for sewage sludge. The 2021 version of the German

Table 6.12 Limits for gas components and odor units in the 2021 German *TA–Luft*

Dust	10 mg/Nm^3 dry
Ammonia (NH3)	20 mg/Nm^3 dry or 0.10 kg/h
Inorganic chloride compounds such as HCl	20 mg/Nm^3 dry or 0.10 kg/h
VOC such as carbon	20 mg C/Nm^3 dry
Odor units	500 OU/Nm^3 wet*

*No odor of raw exhaust gas is allowed to be detected, and at a temperature of 20 °C

TA-Luft (Technical Instructions for keeping air clean) stipulates the following limits for gas component concentrations and odor units of sludge dryers as shown in Table 6.12. It was issued on the 18 August 2021 and is published on the internet under: https://www.verwaltungsvorschriften-im-internet.de/bsvwvbund_18082021_IGI25025005.htm and https://www.bmuv.de/fileadmin/Daten_BMU/Download_PDF/Glaeserne_Gesetze/19._Lp/ta_luft_neu/Entwurf/ta_luft_neu_refe_bf.pdf.

The odor leaving the biofilter must not create an odor impression of sludge drying. Regulations vary from country to country; sometimes, only odor units are limited and the main focus is on odor control. In other cases, specific limits for individual gas components are specified and must be observed.

It is of the utmost importance to clarify the specific limits for exhaust air of a sludge drying system for a project in a timely manner, as these data determine the scope of the supply of the exhaust air treatment unit and, as a result, the CAPEX and OPEX.

In some countries, there is no special regulation for the permissible limits of individual gas components and odor units in the purified exhaust air of sludge dryers. Exhaust gas limits for incineration processes are not suitable for sludge drying processes, as there is no thermal oxidation in the usual exhaust air treatment. It makes sense to adopt the specific regulations for sludge drying of another country or to specify the limits for the individual project.

6.5 Diffusion of Gas Components into Drying Air

Rising air temperature and thus sludge temperature have an increasing effect on the diffusion of gas components from the wet surface of the sludge bed into drying air.

Fig. 6.3 shows the diffusion coefficient of different gas components into air depending on sludge/air temperature.

The linear temperature dependency of diffusion D of the individual gas components into air is clearly expressed. Rising temperature means increasing mass transfer into the

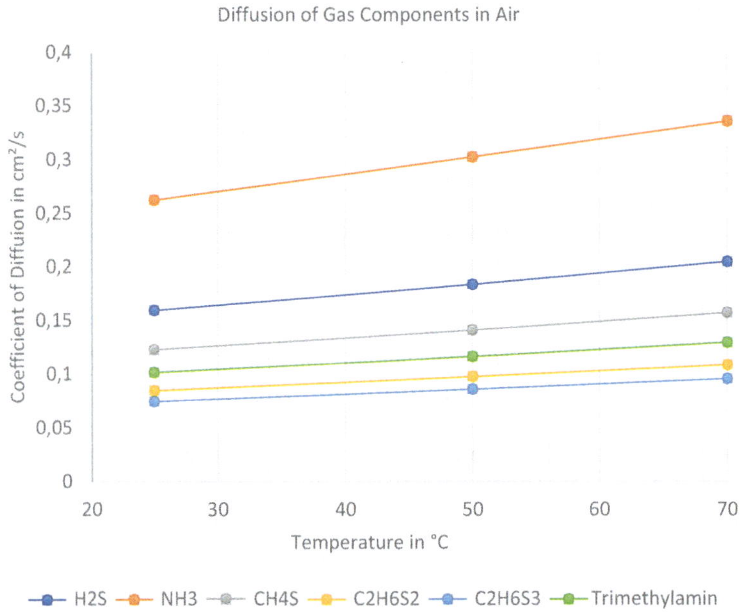

Fig. 6.3 Calculated diffusion coefficients of gas components in air depending on air temperature

drying air. The temperature influence can be expressed by the quotient of temperatures in Kelvin with an exponent of 1.75 (Reid et al. 1977):

$$D \approx \left(\frac{T_1}{T_2}\right)^{1.75} \tag{6.9}$$

6.6 Acid Scrubbers

Scrubbers with chemical additives are commonly used to separate water soluble gases from exhaust air and release them as salts via the regular exhaust water flushing of the scrubber.

For the proper operation of scrubbers, and thus for the entire exhaust air cleaning system, regular cleaning and calibrating of all probes for pH, electrical conductivity, and redox potential (if mounted) is vital. The electrical conductivity of discharged wastewater must be limited to avoid excessive salt deposits resulting in failure of measurements. Also, nozzles spraying recirculated washing liquid in counter- or cross-current flow of exhaust air need to be regularly checked and cleaned.

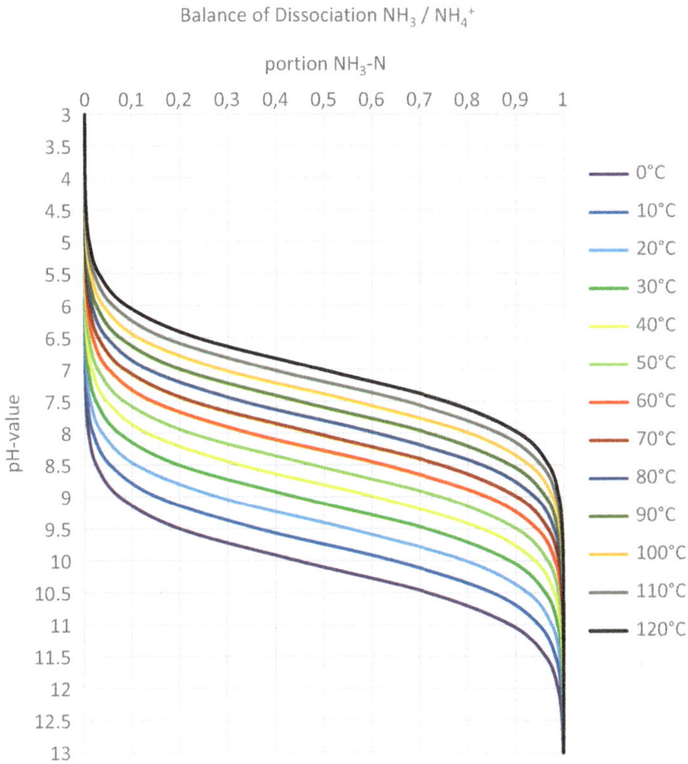

Fig. 6.4 Relationship between ammonia gas (NH_3–N), dissolved ammonium (NH_4^+–N), pH-value, and temperature

Nitrogen compounds with high water solubility such as ammonia and trimethylamine are commonly removed from exhaust air in an acid scrubber. The main gas component in exhaust air is ammonia (NH_3). Fig. 6.4 shows the relationship between ammonia gas (NH_3–N), dissolved ammonium (NH_4–N; in sludge with a wet surface), pH-value, and sludge temperature.

It can be stated that with rising pH-value and increasing sludge temperature, more ammonia (NH_3) will be transferred to the gas phase and consequently into the exhaust air. In the case of a condensing drying system, more ammonium (NH_4–N) will be found in the condensate; a ratio of separation can be assumed to be around 60–70 mass% of the ammonia entering the condenser.

If one assumes a pH-value of 8.2 of a digested sludge according to Fig. 6.3 and a maximum temperature of wet sludge of around 60 °C, one obtains around 40% of the NH_4–N being transferred to drying air as NH_3–N.

Once the ammonia content in the sludge is known—usually covering a wide range of 0.2–2 mass% of DM—the maximum ammonia content in the air can be estimated.

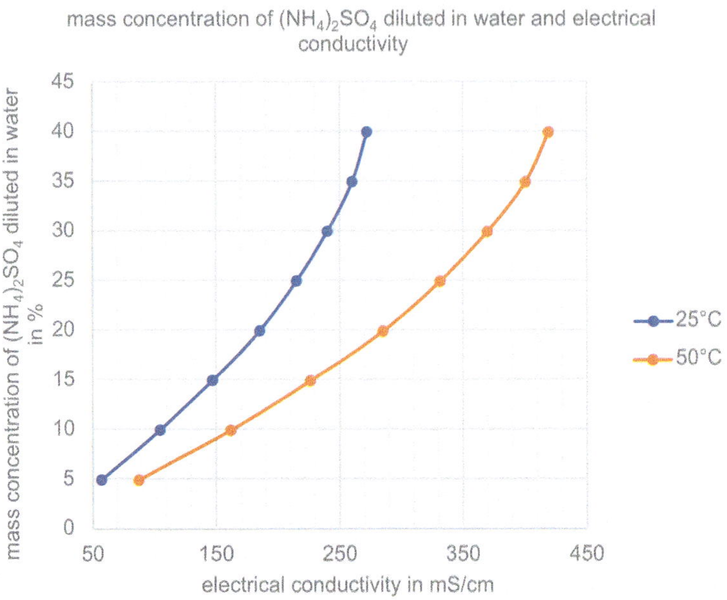

Fig. 6.5 Electrical conductivity and mass concentration of aqueous ammonium sulfate solution in the wastewater of an acid scrubber for a temperature of 25 °C (according to data from Wetter and Thiel 2018) and 50 °C

Formula (6.10) expresses how the different parameters are interconnected:

$$NH_3{-}N = NH_4{-}N \times \frac{10^{pH}}{\exp\left(\frac{6.344}{273+\vartheta}\right) + 10^{pH}} \tag{6.10}$$

A sludge with a DR content of 23%, a pH of 8.2, and an ammonium (NH_4–N) content of 1.5 mass% will generate a maximum of 6.9 g NH_3/kg of DM-throughput during drying. A typical range of values of NH_3 production encountered in belt drying of digested sludge at a drying temperature of ≈ 120–130 °C is between 4 and 9 g NH_3/kg DM.

Sulfuric acid (H_2SO_4) is used to maintain the pH-value in the liquid phase at around 3 during operation. Normally, the efficiency of ammonia separation reaches more than 90–95%, and the minimum ammonia content in exhaust air having passed the acid scrubber is around 1 mg/m^3. The wastewater of the acid scrubber is periodically flushed off, usually when a limit of mass concentration of ammonium sulfate is reached in the sump water. The value is usually set to less than 20 mass%, corresponding to an electrical conductivity of less than 185 mS/cm at 25 °C. The dependency of electrical conductivity and concentration of ammonium sulfate in wastewater is shown in Fig. 6.5.

The dependency of mass concentration of an aqueous ammonium sulfate and the electrical conductivity is expressed by the formula at 25 °C (Wetter and Thiel 2018):

$$C_{(NH_4)_2SO_4}(\text{mass\%}) = 3 \times 10^{-6} \times \kappa^3 - 0.001 \times \kappa^2 + 0.2185 \times \kappa - 4.8434 \tag{6.11}$$

Table 6.13 Crystallization concentration of aqueous ammonium sulfate solution depending on temperature. (Rant 1977)

Temperature in °C	Approximate mass concentration of aqueous ammonium sulfate solution for crystallization in %
0	41
50	45
75	47.7
100	51

with a coefficient of determination of $R^2 = 0.9988$ and κ being the electrical conductivity in mS/cm.

However, operating scrubbers at higher temperatures will increase the electrical conductivity of aqueous solutions. At an equal concentration of NH_4^+, the increase is calculated with a quotient of ion equivalent conductivities of higher temperature to the value of 25 °C. This quotient is 1.543 at 50 °C for NH_4^+. Thus, the increase in electrical conductivity at 50 °C is calculated by multiplying $\kappa_{25\,°C}$ with a factor of 1.543 (Wagner 2012). Practical measurements at 40–45 °C showed an electrical conductivity of aqueous ammonium sulfate solution of 350–400 mS/cm, when the mass concentration approaches 30%.

In any event, the temperature-dependent crystallization of ammonium sulfate is another limit for the concentration in the discharged wastewater. Table 6.13 presents the concentration limits for crystallization of the ammonium sulfate solution depending on temperature.

It must be emphasized that the concentration of ammonium sulfate solution varies at different points of the acid scrubber. Deposits of crystals can often be observed at the spraying nozzles, at the drop separator, and in the sump of the scrubber near measuring probes for pH and electrical conductivity. The formation of deposits is increased with increasing temperature in the scrubber. Therefore, a safe distance to the admissible mass concentration must be applied to avoid failure of the scrubbing system due to deposits. As a rule, the concentration of ammonium sulfate in wastewater discharged from the scrubber is kept below 15–20 mass%. Fig. 6.6 shows salt deposits on a spraying nozzle of an acid scrubber.

A two-stage system of acid scrubber—feeding wastewater from stage 1 to stage 2—can reportedly achieve a higher concentration of $(NH_4)_2SO_4$ at the end of the second stage, but generates higher costs in terms of investment, operation, and maintenance. In the second stage, the feeding of liquid must be carried out with chutes instead of nozzles to avoid blocking by crystallized ammonium sulfate.

It must also be mentioned that dust is separated from raw exhaust air of the dryer in the acid scrubber and that the minimum wastewater discharged is determined by a maximum dust agglomeration in wastewater of around 1 g/L to avoid blocking of the packed bed in the scrubber and to ensure problem-free operation for a longer period.

Fig. 6.7 illustrates the wastewater of an acid scrubber of a fresh air–waste air belt dryer taken in January.

Measuring the electrical conductivity of wastewater from the acid scrubber at a given temperature will indicate the mass concentration of ammonium sulfate in wastewater in

Fig. 6.6 Deposits of ammonium sulfate crystals on a spraying nozzle of an acid scrubber. (HUBER 2019a)

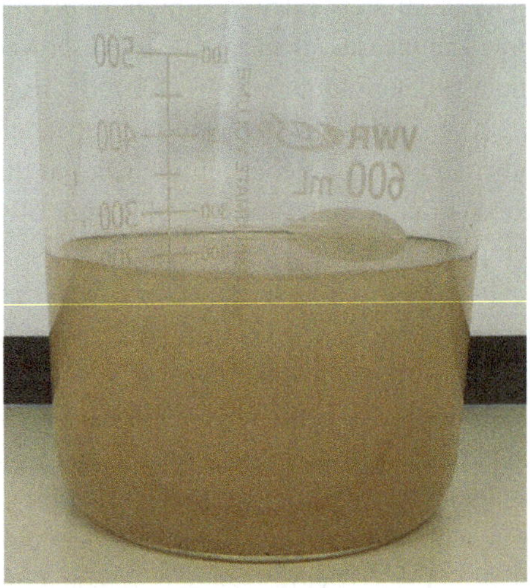

Fig. 6.7 Wastewater from an acid scrubber of a fresh air–exhaust air belt dryer, drying sludge from external sources at an air temperature of 84 °C. (HUBER 2019b)

mass%, assuming that the distribution of the salt is homogenous and the point of measurement is representative. Typical electrical conductivities of wastewater in the sump of an acid scrubber range from 150 to 250 mS/cm, depending on the temperature of the water circulated in the system.

Around 4 kg of sulfuric acid H_2SO_4 is needed to bind 1 kg of ammonia (NH_3) as $(NH_4)_2SO_4$ salt in the liquid phase. This figure includes surplus dosing to compensate for incomplete or parasitic reactions in the system. The chemical reaction is expressed as:

$$2\,NH_3 + H_2SO_4 \Leftrightarrow (NH_4)_2SO_4 \tag{6.12}$$

The efficiency of ammonia extraction depends on the initial load in the exhaust air and the dosing of sulfuric acid. The lower the concentration, the lower the efficiency of separation of ammonia gas. Well-operated scrubbers may reach an efficiency of more than 90–95% at an NH_3 concentration above 100 mg NH_3/m^3 of exhaust air. At higher temperatures of scrubber water, more packed bed volume needs to be applied to compensate for the loss of solubility of ammonia. Acid dosing is controlled by the pH-value of the swamp water; as soon as the pH exceeds 3, new acid is dosed. The scrubber water is released when its electrical conductivity exceeds a limit corresponding to an adjustable limit concentration of ammonia sulfate. In any event, crystallization of $(NH_4)_2SO_4$ must be avoided, as pH and conductivity probes are covered with salt crystals and salt deposits might block the swamp system of the scrubber. A sulfuric acid with a concentration of 78 or 96 mass% is usually used. The filling material and the scrubber container are made of polypropylene (PP) or fiber-reinforced plastic, and a maximum temperature of operation of 80 °C is allowed. The typical nonlinear time of gas residence in the packed bed is around 1–2 s.

Sulfuric acid (H_2SO_4) is used for acid scrubbers for the following reasons:

- Low cost and high availability
- Formation of $(NH_4)_2SO_4$, a salt that can be used as fertilizer after concentration to > 38 mass%, for example, by vacuum evaporation at 80 °C and 300 mbar(a).

Due to its high content of ammonium sulfate [$(NH_4)_2SO_4$], wastewater from an acid scrubber operated with H_2SO_4 can be classified as at least slightly hazardous for water. This would correspond to a water hazard class WGK1 in Germany. Local regulations must be observed, especially with regard to tanks for intermediate storage of wastewater and for the installation of the acid scrubbers, which need a collecting basin in case of leakage. The same classification can be made with sulfuric acid (96 mass%).

H_3PO_4 may be also be used as an acid, but is far more expensive. However, when using a two-stage acid scrubber operated at different pH-values, it is possible to produce a di-ammonium phosphate solution [$(NH_4)_2HPO_4$]. This fertilizer is more expensive and can be used throughout the year, but the economics depend on the local price of H_3PO_4, additional costs of operation, and the price of the liquid fertilizer paid on the local market. The pH-value in the two stages must be precisely controlled to avoid ammonia slack or the formation of fine salts (Lange 1971). Di-ammonium salt is removed from the first stage, and H_3PO_4 is fed into the second stage arranged on top of the first stage. Exhaust air is transferred from the first to the second stage in counter-current flow. Fig. 6.8 illustrates a two-stage scrubber of this kind.

First stage at pH > 5.8–6.0 at 50 °C: di-ammonium phosphate production at a mole ratio of NH_3:H_3PO_4 < 1.5

$$NH_3 + NH_4H_2PO_4 \Leftrightarrow (NH_4)_2HPO_4 \tag{6.13}$$

Second stage at 3.5 < pH < 5.5 at 50 °C: mono-ammonium phosphate production at a mole ratio of NH_3:H_3PO_4 > 1.5

$$NH_3 + H_3PO_4 \Leftrightarrow NH_4H_2PO_4 \tag{6.14}$$

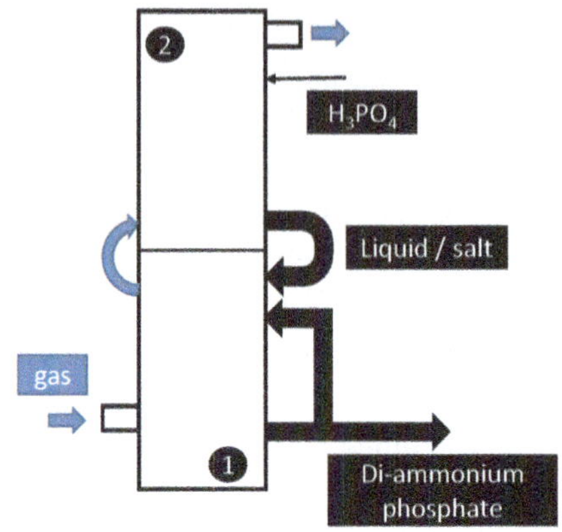

Fig. 6.8 Two-stage scrubber for the production of di-ammonium phosphate. (According to Lange 1971)

The composition of wastewater discharged from an acid scrubber operating with sulfuric acid (H_2SO_4) varies and depends on sludge qualities, pH-value of the sludge, sludge temperature, and water solubility of amino gas components transferred to exhaust air of the dryer, dust in exhaust air, and admissible concentration of $(NH_4)_2SO_4$. The dryer system has an impact, i.e., a dryer with a condensing unit and fresh air–exhaust air drying system. There is some separation of ammonia, dust, and phosphorus by the condensing system, thereby reducing values in the subsequent acid scrubber. Typical values of components in

Table 6.14 Typical values of wastewater of acid scrubbers

Quality of wastewater flushed from acid scrubbers	From scrubber after condensing systems (indirect)	Fresh air–exhaust air unit from scrubber without condensing system
Odor	Uric-like	Uric-like
Color	Yellow to light brown	Yellow to light brown
Chemical oxygen demand (COD) in mg/L	40–**100**–400	≈ 750
Total organic carbon (TOC) in mg/L	≈ 1/3 of COD	Not analyzed
pH	3.0–3.3	2.6–3.6
Conductivity at 25 °C in mS/cm	≈ 50–70*	≈ 130–175
Sulfate (SO_4^{2-}) in mg/L	4000–5000	14,000–20,000
Calculated ammonium sulfate [$(NH_4)_2SO_4$] in mg/L	≈ 5500–6900*	≈ 19,260–27,500
Free sulfuric acid (H_2SO_4), corresponding to SO_4 not bound to NH_4–N in mass%	≈ 0.04	Not analyzed
Acid capacity pH = 4.3 in mmol/L	< 0.1	Not analyzed
Phosphorus in mg/L	0.02–0.50	Not analyzed
Filterable matter (0.7 µm) in mg/L	≈ 200	≈ 700
Chlorine (Cl^-)in mg/L	2–15	Not analyzed

*Typically around 60–70% of the ammonia charge of the exhaust air is removed in the condensing device

wastewater of an acid scrubber are given in Table 6.14; however, it must be emphasized that these values vary according to the type of sludge, its constituents, and pH-value as well as sludge stabilization, intermediate storage of the sludge, mechanical handling, and drying temperature.

6.7 Neutral Scrubbers

If only water is used in the scrubber, the efficiency of ammonia separation drops to less than 60–70%—a value that can also be assumed for water-based condensing units—and the scrubber will become blocked after a certain period of time in the case of higher hydrogen carbonate-contents (HCO_3^-) by carbonate deposits (calcites). In the areas below of the three curves in Fig. 6.9, carbonate deposits such as Na_2CO_3, $MgCO_3$, and $CaCO_3$ are formed and the speed of formation increases with increasing pH-value and hydrogen carbonate content of scrubber water. The relationship between concentrations of CO_2, HCO_3^-, and pH-value in water is shown in Fig. 6.9 (according to Worch 2015). In addition, there is also the danger of ammonia returning to the gas phase when the pH-value is changed. For these reasons, operation of the scrubber without sulfuric acid cannot be recommended. Even hardness-reducing agents do not prevent carbonate deposits in the scrubber for a prolonged period and they need to be frequently removed. The straight lines are for constant pH-values.

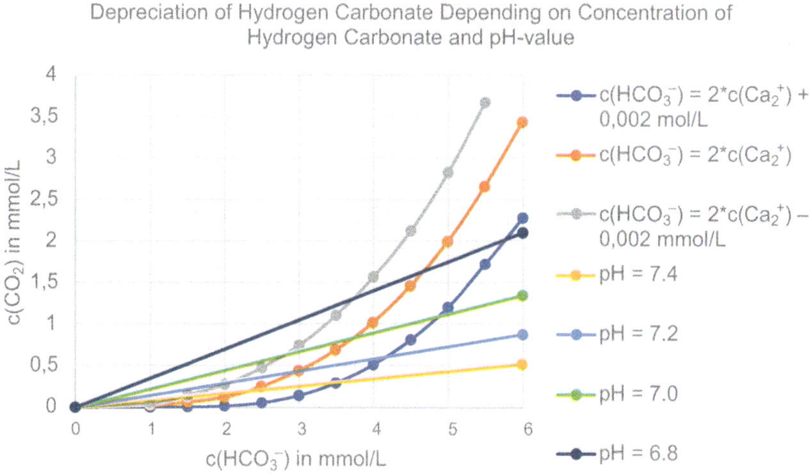

Fig. 6.9 Tillmans curve, calculated for $c(HCO_3^-) = 2\,c(Ca^{2+})$ at 25 °C and for two other equivalent concentration differences. The straight lines describe the equilibrium between CO_2 and HCO_3^--concentations in wastewater for constant pH-values. A wastewater, whose composition results in a pair of values on the curves is in the carbonate—carbon dioxide—equilibrium. Above these curves water is carbonate dissolving, below carbonate precipitating. (Modified from Worch 2015)

6.8 Caustic or Alkaline Scrubbers

In the case of the alkaline (caustic) scrubber, chemicals such as hydrogen peroxide (H_2O_2) and sodium hydroxide (NaOH) are used to form salts from sulfur-based gas components and volatile carbon acids. An even stronger oxidation agent is NaClO, sodium hypochlorite preventing resolution of formed salt. However, temperatures exceeding 60 °C result in increasing destruction of NaClO, thereby considerably reducing its chemical efficiency. Due to the low contact time of less than 3–5 s of exhaust air in the mass transfer zone of the scrubber, the formation of gaseous chlorine components can be neglected.

The pH-value is adjusted between 9.5 and 10.3 by adding of NaOH.

The second chemical agent, hydrogen peroxide (H_2O_2), is optionally controlled by a probe measuring the REDOX potential of sump water, and the limit value for dosing H_2O_2 typically ranges from 90 to 120 mV (Trenkwalder 2011). As a rule, sump water is flushed off at ≥ 25–35 mS/cm of electrical conductivity and an almost equal quantity of fresh water is filled. Refilled fresh water needs to be descaled to a concentration of $(1/2\,Ca^{2+} + 1/2\,Mg^{2+})$ of less than, for example, 0.01 mmol/L to avoid additional consumption of NaOH resulting from more frequent discharging of sump water due to higher electrical conductivity caused by the ions.

Hydrogen sulfide (H_2S) and methane thiol (CH_4S) are separated in a caustic scrubber using NaOH and H_2O_2 as chemicals. Salts are formed and removed from the system by wastewater periodically flushed off. Hydrogen sulfide reacts according to formula (6.15):

$$H_2S + 2\,NaOH + 4\,H_2O_2 \Leftrightarrow Na_2SO_4 + 6\,H_2O \tag{6.15}$$

Thus, for the removal of 1 kg H_2S, 5.96 kg of NaOH (40 mass%) and 13.30 kg H_2O_2 (30 mass%) are theoretically required. Methane thiol is transformed according to formulas (6.16 and 6.17):

$$CH_4S + NaOH \Leftrightarrow CH_3SNa + H_2O \tag{6.16}$$

$$CH_3SNa + 3\,H_2O_2 \Leftrightarrow CH_3SO_3Na + 3\,H_2O \tag{6.17}$$

In the case of 1 kg CH_4S, 2.06 kg of NaOH and 7.10 kg of H_2O_2 are, in theory, required for removal.

However, there are parasitic reactions between NaOH and CO_2 of the air dissolved in water, consuming the dosed chemical agent NaOH:

$$CO_2 + NaOH \Leftrightarrow NaHCO_3 \tag{6.18}$$

and

$$CO_2 + 2\,NaOH \Leftrightarrow Na_2CO_3 + H_2O \tag{6.19}$$

The source of carbon dioxide is not only the natural content in ambient air, which usually contains approximately 420 ppm at the time of publication of this book. CO_2 is also released from dewatered sludge during the drying process. From a measurement of CO_2 in the exhaust air of a fresh air–exhaust air drying system operating at 75 °C air temperature, the specific CO_2 contribution of a digested sludge was calculated to be approximately 40 g/kg of DM fed to the dryer. Only a portion, usually less than 10 mass% of the incoming CO_2, reacts with NaOH. Nevertheless, consumption of NaOH will be considerably higher. For the safe separation of sulfuric gas compounds, a surplus consumption of more than 60% of the chemical agent in comparison to theoretical values for reactions with H_2S and CH_4S needs to be considered.

As already mentioned, sodium hypochlorite (NaClO) is sometimes used as a very strong oxidation agent instead of H_2O_2, but the temperature in the scrubber must be limited to less than 60 °C to avoid the decomposition of NaClO before a chemical reaction with odorous gases can occur. Due to the short reaction time in the caustic scrubber of usually less than 3–5 s, the formation of chlorine gases is not expected. In any event, amines that occur, such as NH_3, must be removed in an acid scrubber to avoid formation of amine chlorites in the caustic scrubber.

Typically, the separation of VOC in a caustic scrubber is less than 15–20 mass% of VOC entering the scrubber for a fresh air–exhaust air system. In the case of a condensing system, where a portion of VOCs is separated in the condenser, the degree of additional

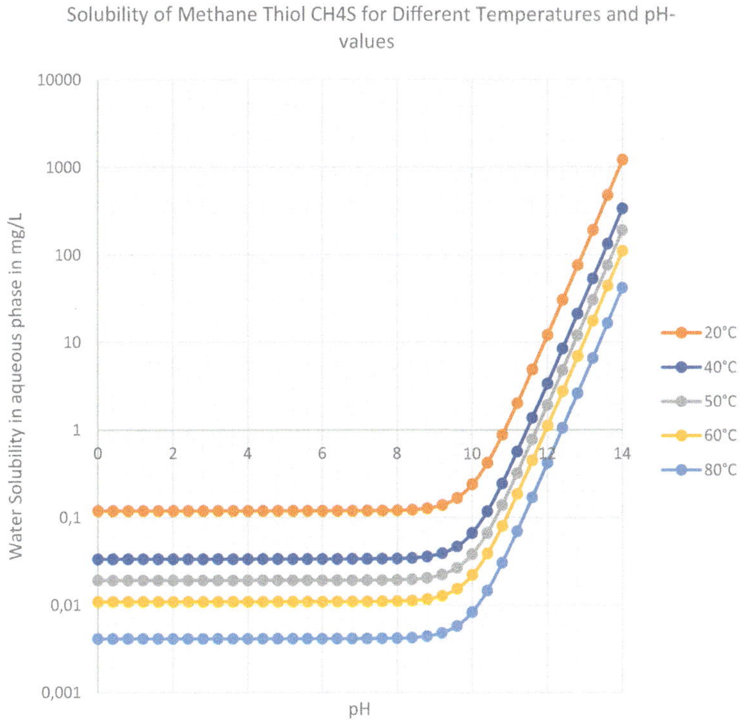

Fig. 6.10 Solubility of methane thiol depending on pH-value and temperature. (Michaud 2022; Zin 2017)

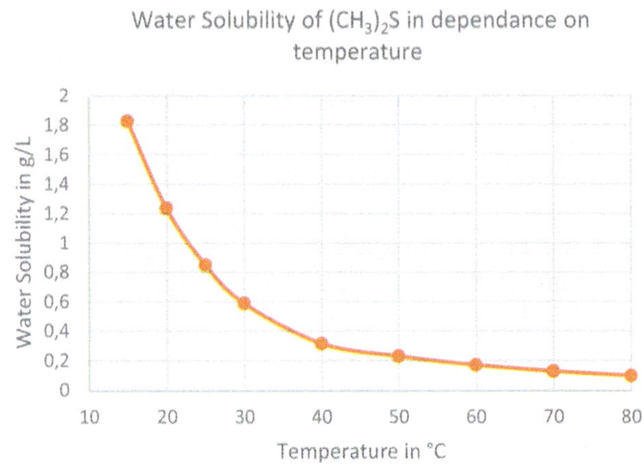

Fig. 6.11 Water solubility of $(CH_3)_2S$ depending on temperature

removal ranges up to 8%. Exceptions might be encountered in the case of a high portion of volatile carbon acids with high water solubility. High methane thiol contents afford a higher pH-value > 10.3, as the water solubility of CH_4S increases considerably in the region of higher alkalinity (De Bruyn et al. 1995). However, parasitic reaction with CO_2 also increases at higher pH-values. Fig. 6.10 illustrates the solubility of CH_4S in water depending on pH-value and temperature (Michaud 2022; Zin 2017).

In contrast to CH_4S, dimethyl sulfide [$(CH_3)_2S$] has a very low rate of hydrolysis and as a consequence, there is in practice no dependence of water solubility on pH. The influence of the temperature of liquid in the scrubber on water solubility of $(CH_3)_2S$ is expressed in Fig. 6.11. The calculation has been made with a vapor pressure estimation based on the Antoine equation (Yaws 2015). The graph demonstrates that there is a huge decline of water solubility in the temperature range from 15 to 40 °C.

Typical values of wastewater from caustic scrubbers are given in Table 6.15.

Table 6.15 Typical values of wastewater flushed off from caustic (alkaline) scrubbers

Quality of wastewater flushed off from caustic scrubbers	From scrubber after acid scrubber and after condensing systems (indirect)
Chemical oxygen demand (COD) in mg/L	30–80
Total organic carbon (TOC) in mg/L	≈ 1/3 of COD
pH	10.0–10.6
Conductivity at 25 °C in mS/cm	3000–4000
Sodium (Na) in mg/L	1200–1600
Hydrogen carbonate (HCO_3^-) in mg/L	2800
Acid capacity pH = 4.3 in mmol/L	40–50
Phosphorus in mg/L	< 0.02
Filterable matter (0.7 µm) in mg/L	≈ 0.50
Chlorine (Cl^-) in mg/L	< 2

Due to its content of Na_2SO_4 and other sodium salts, wastewater from a caustic scrubber operated with NaOH can be classified as at least "slightly hazardous for water." This would correspond to a water hazard class WGK1 in Germany. The same classification can be made with caustic soda NaOH (50 mass%) and hydrogen peroxide H_2O_2 (30 mass%). Sodium hypochlorite is a chemical agent classified in the higher hazard class of WGK2, i.e., "remarkably hazardous to water." All tanks must be equipped with a leakage basin underneath. Local regulations for the storage of chemical agents must be observed.

6.9 Typical Design of Chemical Counter-Current and Cross-Flow Scrubbers

A standard exhaust air treatment system of a belt dryer with a condensing line consists of an acid scrubber, an alkaline scrubber, and a biofilter. This exhaust air treatment system is normally used for drying the sludge of a wastewater treatment plant. Fig. 6.12 shows a typical arrangement of a counter-current pair of scrubbers (acid and caustic).

Typical technical data and features of chemical counter-current scrubbers are listed below:

- Specific airflow for empty column: ≈ max. 2.7 $m^3/(m^2$ of scrubber cross section × s). The gas load factor F is calculated by: $F = w \times \sqrt{\rho_{gas}}$. Its value is typically around 1–3 $Pa^{0.5}$ (VDI 3679 1999/2005). Another source stipulates a specific airflow of 1.5–2.5 m/s (WEF 2020).
- Typical air resistance (static): ≈ 600 Pa (with geometric demister, without wire demister).
- Typical specific scrubber water circulation: 15–25 $m^3/(m^2$ of scrubber cross section × h). Typically 12 $m^3/(m^2 \times h)$ or 5 GPM/ft^2 are indicated in another source (WEF 2020). In the case of gas components with low solubility in water: Higher specific water circulation must be chosen. Values of up to 80 $m^3/(m^2 \times h)$ are applied (VDI 3679 1999/2005).
- Typical linear residence time of air in a packed bed: ≈ 1–2 s for acid scrubbers, ≈ 2–4 s for caustic scrubbers.
- Typical linear residence time of water in a packed bed: ≈ 8–15 min
- Typical exhaust air load on a packed bed: max. ≈ 3350 m^3 of exhaust air/(m^3 of a packed bed × h) and 7700–9000 m^3 of exhaust air/(m^2 inflow cross section of a packed bed × h). Mass transfer increases with gas speed proportionally $w_{gas}^{0.6}$. Gas/air flow is limited by the flooding effect, describing the fact that scrubber water is no longer able to flow down by gravity, but is dammed up.
- Typical packing material for low resistance and low liability to scaling (see Fig. 6.13): 100 mm (4 inch) rectangular body, porosity 96%, specific weight ≈ 45 kg/m^3, specific surface ≈ 98 m^2/m^3, material glass fiber reinforced PP with maximum temperature for continuous operation of 110 °C or 2 inch round body, porosity ≈ 148 m^2/m^3. The

Fig. 6.12 A pair of counter-current scrubbers (acid and caustic) for the first stage of exhaust air cleaning of a belt dryer, installed with measurement of level, electrical conductivity, and pH-value of sump water. (Heindl 2020)

packed bed material and structure are chosen depending on dust load in exhaust air and liability of packed bed to become blocked by deposits.

- Typical nozzles: up to six pieces of 1 1/2 inch, AG 120° spraying angle, or one big bulb nozzle with less liability to blocking by salt deposits.
- Droplet separators (demister units)—either wave-plate separators or regular packings—with a height of 20–25 cm and draining/collecting device for droplets for retransfer to the packed bed, design for droplet diameter separation limit < 30–$40\,\mu m$, air resistance ≈ 80–$100\,Pa$. The best droplet separation and lowest limit for droplet diameter is achieved at almost maximum speed, typically being around 4–5 m/s. Droplet diameter limit can be reduced to less than 5–10 µm by additionally using wire filters of 10–15 cm height, made of plastic or stainless steel wires. Thus, scrubber water transfer to subsequent exhaust air treatment units is efficiently reduced. However, blocking of the demister unit caused by crystallization of captured droplets must be prevented by regular spray cleaning. Additional air resistance of 200–300 Pa at 4 m/s needs to be accepted (Bürkholz 1989).
- To reduce the wall effect in exhaust air treatment: an inner diameter of the scrubber > 10-fold of the diameter of packed bed elements, and height of the packed bed > 1.5 fold of the inner diameter of the scrubber, as a rule of thumb.

Fig. 6.13 shows a packed bed in a scrubber used for exhaust air treatment.

The concentration of scrubber wastewater flushed off—in the case of an acid scrubber, for example, the concentration of liquid $(NH_4)_2SO_4$—is limited by the formation of

Fig. 6.13 A packed bed in a scrubber for exhaust air treatment. (HUBER 2020/22)

salt deposits occurring at probes, nozzles, droplet separators, or in the scrubber sump. For trouble-free long-term operation, salt deposits must be limited and the electrical conductivity of salt solution set correspondingly. In addition, it is also recommended to limit agglomeration of dust inside the scrubber. The fresh water supply corresponding to wastewater discharge must be adjusted to limit dust content in wastewater to less than 1 mg/L.

Special attention must be paid to the design of the droplet separator arranged in the headpiece of the scrubber. Carry-over of droplets with high salt content must be avoided as far as possible, as deposits of salt crystals may block subsequent heat exchangers or activated carbon filters. Frequent cleaning of the droplet separator by water spraying retains the function of this important device.

Water solubility of most polar gas components decreases considerably with increasing liquid temperature in the scrubber. This needs to be considered in the design of the scrubber; if possible, cooling of scrubber water with partial condensing of water vapor from the dryer exhaust air can be used.

Mass transfer in packed bed counter-current reactors increases with exhaust air flow proportional to $w_{\text{gas}}^{0.6}$ (Schmidt 1970). This corresponds to an increase of approximately 50% by doubling gas velocity upwards. However, gas resistance also increases, and the droplet separator needs to be more efficient to avoid tramp droplets forming deposits in subsequent elements of the exhaust air treatment unit.

Crossflow scrubbers are sometimes used and the exhaust air is conducted horizontally through a packed bed, whereas the scrubber water flows downwards. The crossflow scrubbers have a higher demand in terms of footprint and a lower mass transfer efficiency for a given length of packed bed. For equal separation performance, the length of the packed bed in the direction of exhaust air flow will be around 40% larger in comparison to counter-current flow scrubbers. However, crossflow scrubbers have lower demand in terms of installation height and better accessibility for maintenance and cleaning works. Fig. 6.14 shows a crossflow scrubber with a shorter stage for ammonia separation and two

Fig. 6.14 Three-stage crossflow scrubber for exhaust air treatment; the *red motors* are for scrubber water circulation. (Tholander 2022)

stages for caustic separation of sulfur components (Tholander 2022). The length of packed beds is nearly three times greater for the two-stage caustic scrubber part in comparison to the one-stage acid scrubber part, corresponding to a longer retention time. The two-stage design of the caustic scrubber prevents back-mixing of scrubber water to a great extent in comparison to one extended stage.

6.10 Trickle Bed Reactors and Bioscrubbers

Bioscrubbers and trickle bed reactors are wastewater-generating units for exhaust air treatment. This is especially the case with activated sludge tanks. Thus, at sites with limited or insufficient wastewater treatment, these bioreactors cannot be used. However, this is not an issue at wastewater treatment plants. The limited adaptability of microorganisms to fluctuating exhaust gas composition—normally encountered in projects involving the drying of sludge from external sources—restricts their use. A typical microorganism that can be used for inoculation is *Pseudomonas Putida*, which is able to oxidize a large number of volatile compounds. After a certain time of operation, however, a mixed population of microorganisms will develop (Rutenfranz 1996).

Packed bed height ranges from 1 to 5 m for bioscrubbers and 2–10 m for trickle bed reactors. Typically, the pH-value ranges from 5.5 to 9 and the temperature from 15 to 40 °C. The bed porosity is around 50% for lava stones and around 95% for specific fillers (WEF 2020).

Trickle bed reactors consist of one tank where the gas–liquid mass transfer and the biological degradation of volatile matter is carried out in one apparatus. The airflow of crude gas is downwards, and liquid is sprayed on the packed bed in concurrent flow. The specific liquid density ranges from 5 to 20 $m^3/(m^2$ of cross section × h). The retention time of exhaust air must be up to 10 s for components with good water solubility and up to 30 s for gas components with fair water solubility. Exhaust gas flow ranges from 100 to

Fig. 6.15 A bioscrubber used to reduce NH_3. (Likusta 2022)

$1000\,m^3/(m^2 \times h)$ (VDI 3478 1996). Another source mentions around $1500\,m^3/(m^2 \times h)$ (WEF 2020).

Bioscrubbers consist of an absorber unit with a packed bed and an activated sludge tank arranged underneath or to the side. Typically, airflow is directed upwards and liquid flow in the counter-current direction. For gas components of good biological degradability, stable operation can be achieved as long as the volume load does not exceed 0.4 kg chemical oxygen demand (COD)/(m^3 of packed bed × h) (VDI 3478 1996). Liquid flow density for packed column scrubbers ranges from 20 to $60\,m^3/(m^2$ of cross section × h).

Bioscrubbers are sometimes used for exhaust air treatment to remove ammonia. Biofilms are created on the surface of fillers or filling racks. The biomass concentration ranges from 5 to 8 g/L, and gas velocities are set to 1–3 m/s upwards in packed towers. Economical application in VOC removal is restricted according to Shareefdeen (Shareefdeen and Singh 2005) to substances with a non-dimensional Henry coefficient H of less than 0.01 (e.g., acetone $H = 0.0016$ or n-butanol $H = 0.00035$), i.e., good water solubility. Higher Henry coefficients of up to $H = 5$, however, are also mentioned in the literature (Shareefdeen and Singh 2005). When cleaning or maintaining these scrubbers, one must bear in mind that a portion of the ammonia is transferred by microbes to NO or NO_2, gases that are harmful to humans.

In combination with a preceding physical scrubber, a bioscrubber may be useful in reducing phenols, sulfur compounds, organic acids and their esters, aldehydes, and alcohols. The nutrients necessary for bacteria growth as well as chemicals for adjusting pH-values need to be dosed to optimize efficiency. Installing a bioscrubber in front of a biofilter may relieve the final stage of exhaust air cleaning with regard to VOC, thus reducing biofilter size. Fig. 6.15 shows a bioscrubber (Likusta 2022).

6.11 Biofilters

Mesophilic biofilters are usually used for final exhaust air cleaning. These filters are also suitable for the degradation of poorly or non-water soluble VOC components. Their great advantages include the adaption of the microbial population to the exhaust air components and the long contact time of air to biofilm of at least 45 s. It must be emphasized that biofilters require a certain start-up period—usually 2–3 weeks—to adapt the microbial population to the "nutritional" value of the exhaust gas. Torn wood roots ($1/4$–$1/5$ of the height of the filter at the bottom) and chips ($3/4$–$4/5$ on the top) are used as basic materials, where microorganisms are able to settle on wet surfaces. Also, fiber peat and bark mulch (grain diameter of > 50 mm) are used. The bed must be replaced after 3–5 years when the basic materials are broken down to fines and air resistance of the filter considerably increases. Specific costs of around € 30–€ 40/m^3 need to be calculated for replacement (2015 price basis). The filter must be kept wet at all times to ensure optimum growing conditions for microorganisms; therefore, a neutral scrubber or a water spraying device are often placed upstream. In the case of a covered biofilter, water pipes are mounted above the bed for spraying. Bio filter produce their own odor, typically in the range of 200–300 OU/m^3.

Methane is not metabolized by microorganisms within a reasonable period of time, resulting in a non-economic size of biofilter.

For satisfactory operation, certain design and operational conditions must be met (Bank 1995; VDI 3477 2004 & newest edition 2016; Reiser and Fischer 1996; Shareefdeen and Singh 2005):

- Material and porosity of the filter bed at a moisture content of 50% relative to total mass—mixture of bark compost and puffed clay: porosity $\approx 35\%$, mixture of fiber peat and heather: 65%, wood chips: $\approx 55\%$, mixture of coconut fiber and fiber peat: $\approx 60\%$. Only coconut fibers are not suitable for biofilter use due to low water-holding capacity, low content of nutrients and salts, and extremely low microbial degradation. A biofilter consisting of wood root parts and wood chips is considered an efficient material mix for a biofilter. Organic materials are suitable to compensate fluctuations in temperature and humidity in the filter bed. Expanded clay or styrofoam beads can be added to the organic filter material to reinforce mechanical stability and water-holding capacity.
- Carbon metabolizing microorganisms need sulfur, nitrogen, and phosphorus to build up their cell structure. The relation of fed mass of C to N to P should be 100 to 5 to 1; high mass degradation of C-loaded gas components entering the biofilter may afford corresponding N and P dosages with the spraying device. For example, urea can be

used as a nutritional agent (around 0.03–0.04 mass% of DM of urea [CH_4N_2O] relative to carbon [C] fed to the biofilter).

- Diameter of bed pores: > 50 µm, at best approximately 100 µm to ensure an even distribution of exhaust air.
- Contact time of exhaust air in the bed, calculated taking into account a bed porosity of usually 0.4–0.6: 40–90 s, typically 45 s. In the case of very high VOC load, the retention time can also be considerably longer.
- Exhaust airflow ratio: approximately 40–100 m^3 of exhaust air per m^3 of biofilter bed and hour; for high performance biofilters, up to 180–200 m^3 exhaust air per m^3 of bed and hour, depending on filter material and concentration of VOCs in the exhaust air (Margesin et al. 1996). The specific surface loading rate typically ranges from 100 to 150 m^3 exhaust air per m^2 and hour.
- The load of VOC in kg/h must also be observed and may reduce specific exhaust airflow, thus resulting in larger biofilters. As a rule of thumb, approximately 1 L of water needs to be supplied to the biofilter per kg of degraded VOC (VDI 3477 2016). High mass flow of VOC transferred to the biofilter may result in earlier degradation of biofilter material and even higher air flow resistance of the filter during operation.
- Height of biofilter bed: approximately 1.0 (3.3 ft)–1.8 m (6 ft), maximum 2.0 m (6.5 ft).
- Air resistance of the biofilter including air distribution: maximum 500–600 Pa/m, whereas the bed itself, when fresh, has a resistance of 200 Pa/m. After 3–5 years, the pressure drop of the bed increases due to mineralization of organic material, and filter material needs to be replaced.
- Air distribution beneath the biofilter in an upstream design must be as even as possible; special distribution units improve the situation.
- Relative humidity of exhaust air entering the biofilter: > 95–98%.
- Ventilators installed in front of a biofilter result in a temperature rise of approximately 1 K per 1000 Pa of pressure increase and reduce the relative humidity of exhaust air entering the biofilter. Water injection needs to be applied to keep relative humidity at values > 95–98% and ranges from 1 to 10 L/m^3 of exhaust air with an airspeed of < 3 m/s. In the case of containerized biofilters with air downflow through the filter bed, ventilators should be placed downstream. The design of the biofilter must be adapted to a vacuum operation of, for example, −3000 to −5000 Pa.
- Moisture of the bed: 40–80%, relative to total mass, optimally 50–60%. Higher contents of lipophilic gas components in exhaust air enable lower water content of the filter material of 50–60%. For practical purposes, manual compression of filter material may not result in the drainage of water. Dried biofilter material leads to the formation of air channels and uneven air distribution, together with reduced microorganism activity. The material bed is wetted from the top by water spraying devices and from the bottom by exhaust air fully saturated with water vapor. One must bear in mind that exhaust air ventilators installed in front of biofilters typically heat up exhaust air by 1–3 K depending on the total pressure difference. Of course, this reduces the relative humidity of exhaust air.

- Maximum temperature of exhaust air at the entrance: 40–42 °C for mesophilic organisms, while for thermophilic microbial populations, the window of application ranges from 45 to 65 °C. The recommended temperature range for mesophilic biofilters is 20–40 °C, optimally 33–37 °C.
- pH-value range: 7 ± 1; higher or lower pH-values, caused by excessive loads of ammonia (NH_3) or sulfuric components such as H_2S in exhaust air, may result in a breakdown of the microorganism population when the buffer capacity of the biofilter material is exhausted. Therefore, admissible concentrations of H_2S and NH_3 are limited to values of less than $5\,mg/m^3$. Carbon dioxide (CO_2) and sulfate (SO_4^{2-}), generated by metabolic activities of microorganisms, tend to reduce the pH-value of the filter. Operation at lower pH-values is reported to be possible, but will result in natural selection of microorganisms.
- The maximum dust concentration of exhaust air entering the biofilter should be limited to less than $< 1\,mg/m^3$, otherwise blocking of the filter bed may occur in long-term operation.
- Wetted parts with very high air resistance as well as air channels in the bed with low bulk density must be avoided to ensure even airflow through the bed as much as possible. Heat is generated by biological reactions and, as a result, a higher load of VOCs speeds up drying of the biofilter and there is increased demand for water. The degradation of 1 kg VOC requires a water supply to the biofilter of around 1 L.
- Increased degradation of VOC in the biofilter can be achieved by using molds with very high pH-tolerance (Franke 2011) as well as by the downstream flow of exhaust air with better control of humidity in the bed.
- The acclimatization period of microorganisms ranges from 10 to 30 days. The survival time of microorganisms in a non-operational biofilter may reach up to three weeks, longer periods will lead to a prolonged time of reactivation of full biofilter capacity. The start-up period is shortened to a few days when specific microorganisms already adapted to the VOC or volatile inorganic compounds (VIC) are inoculated.
- The maximum elimination capacity is reached when an increase in gas concentration of the exhaust airflow no longer results in a rise of volume-related elimination of the gas component. In the case of H_2S as a single gas component, a 40-g H_2S load per m^3 bed an hour equals degradation per m^3 bed and hour. Higher loads cannot be treated without breakthrough of H_2S.
- Biofilters are able to treat less hydrophilic substances with non-dimensional Henry's law constant of less than 10. Methane has an H-value of 27, cyclohexane 7.3, and toluene 0.28 (Shareefdeen and Singh 2005).
- Short-term peaks in the concentration of gas components in exhaust air can be compensated for by biofilters for a period of up to 10 min (Margesin et al. 1996).
- Ozone feeding to exhaust air in front of a biofilter might lead to a stress situation for microorganisms during a start-up phase, but could improve degradation of VOCs in steady-state operation. Alkanes (C_nH_{2n+2}) (methane, ethane, propane, buthane, pentane) are not oxidized by ozone (Dobslaw 2022). Wang reported a positive effect for

ozone feeding to exhaust air on degradation of chlorbenzene (C_6H_5Cl) in a biofilter. This was explained by limited thickness of the biofilm and increased biofilm activity (Wang et al. 2009).

- The staged operation of biofilters in a line improves degradation of different VOCs if the first biofilter is operated at higher moisture contents and the second biofilter at lower moisture contents. Thus, the pH-value can be adjusted.
- Biofilters based on a concrete construction must be provided with sufficient corrosion resistance.
- Weed growth on the surface of the biofilter must be avoided with regard to flow restrictions in the filter bed.
- Wastewater or sump water—resulting from condensation effects in the biofilter—is lower in the case of covered biofilters and may be loaded with odorous substances. Using this water for rewetting the biofilter is only feasible for exhaust air in front of the biofilter and not for surface wetting. Otherwise, the sump water needs to be treated in a wastewater treatment plant. As a rule, the COD content of drainage water is around 1500 mg/L, its pH-value less than 7. Humic acids and hemicellulose are major components in this wastewater. The amount of wastewater can be estimated by assuming a temperature drop in the biofilter of 3 K, i.e., from 40 °C and fully saturated to 37 °C. Approximately 7–7.5 g of water per m^3_{oper} of 40 °C will condensate, depending on total ambient pressure.
- The number of microbiel cells has to be around 107–108 colony founding units (CFU) per gramm of bio filter material for a proper operation of biofilter. This corresponds to around 106–107 CFU/mL of drainage water (Dobslaw 2023).

Table 6.16 shows reduction rates of biofilters for different specific odor-carrying gas components under the conditions that the biofilter is operated at optimum temperatures of around 35–37 °C, deviations of pH-values in the wet phase of filter around 7 are small, and specific air flow is less than 150 m^3 of exhaust air per m^2 of the biofilter and hour, giving a retention time for air of far more than 35 s (Ecochimica 2010). The data are valid under optimal conditions for microorganisms in a biofilter and for sufficient humidity of

Table 6.16 Maximum reduction rates of biofilters for different odor-carrying gas components, operated at suitable conditions with regard to temperature, humidity, and pH-value in liquid phase. (Ecochimica 2010)

Gas component	Maximum reduction rate in % of gas mass entering the biofilter
VOC in general	Up to 98.9
Nitrogen compounds	Up to 97.9
Aldehydes	92–99.8
Volatile acids (organic acids)	Up to 99.9
Mercaptans	92–95
Hydrogen sulfide (H_2S)	98–100
Ammonium (NH_3)	92–95

the filter ensured by a neutral scrubber in front of the filter and additional water spraying of the filter surface exposed to the surrounding environment. In addition, the aeration system must ensure optimum distribution of exhaust air in the filter bed.

As a rule, the degradation efficiencies of laboratory tests need to be reduced to 2/3 for use in industrial plants (Dobslaw 2022). It must be emphasized that, depending on the type of VOC, the reduction efficiency of biofilters differs considerably. Alcohols, aldehydes, ketones, carboxylic acids, and carboxylic acid esters are metabolized comparatively well with a ratio of 90–95%. Organic sulfides such as thioethers and aromatic hydrocarbons (toluene, xylene, benzene, styrene) are removed by less than 50% of the feeding mass in normal biofilters. Under the condition of sufficient linear retention time of exhaust gas in the biofilter of more than 70 s, around up to 70% of volatile sulfuric VOCs such as dimethyl sulfide or dimethyl disulfide are removed by microbial action. As a rule of thumb, C11–C18 compounds are degraded by 70% in biofilters (Dobslaw 2022). It is well known that saturated aliphatic hydrocarbons such as methane, pentane, or hexane cannot be sufficiently reduced in normally designed biofilters with a 45-s retention time. Very slow degradability is encountered in the case of halogenated and polyaromatic hydrocarbons. Linear hydrocarbons, even with very long chains such as tridecane or tetradecane, can be efficiently degraded in strains of C4 as long as they can be accessed by special degrading enzymes. Cycled or branched compounds of hydrocarbons or amines show higher resistivity to degradation by microorganisms in a biofilter. With the exception of N_2O, VICs such as H_2S, NO_x, SO_2, HCl, NH_3, PH_3, SiH_4, and HF are rapidly degraded (Shareefdeen and Singh 2005). Long-chain silicone compounds cannot be efficiently metabolized in biofilters, but the filtering feature of the biofilter also results in a reduction of these compounds. Concerning the reduction of odor units, biofilters are able to achieve up to 90–95%; however, the initial load should not exceed 15,000 to 20,000 odor units per m^3 of exhaust air for the general design of a 45-s retention time of exhaust air in the filter bed (Dobslaw 2022). Up to 50,000 odor units may be acceptable with adapted, i.e., increased, biofilter volume and increased retention time of exhaust air in the biofilter.

In higher concentrations, some hydrocarbons may be toxic to microorganisms such as phenols or saturated hydrocarbons like undecane (Schweisfurth 1988). SO_2 is also harmful to microorganisms (VDI 3477 2016). For these substances to become decomposed, a certain minimum time of adoption is necessary and sufficient oxygen needs to be made available. Gaseous hydrocarbons must be transferred via the liquid phase, where nutrients are stored, to the solid phase, where microorganisms settle.

High concentrations of aerosols, fat as well as dust may lead to blockages of the filter bed and need to be removed in an upstream device.

In the special cases of predominantly odorous gas components, the biofilter must be inoculated with specialized microorganisms. There are microorganisms specialized for the decomposition of sulfur gas components with very low water solubility such as dimethyl sulfide in the transition from mesophilic to thermophilic temperature conditions (40–50 °C), namely *Pseudonacardia asaccharolytica* and *Pseudonocardia sulfidoxydans sp.* (Reichert et al. 1998). An image of *Pseudonacardia asaccharolytica* is shown in Fig. 6.16.

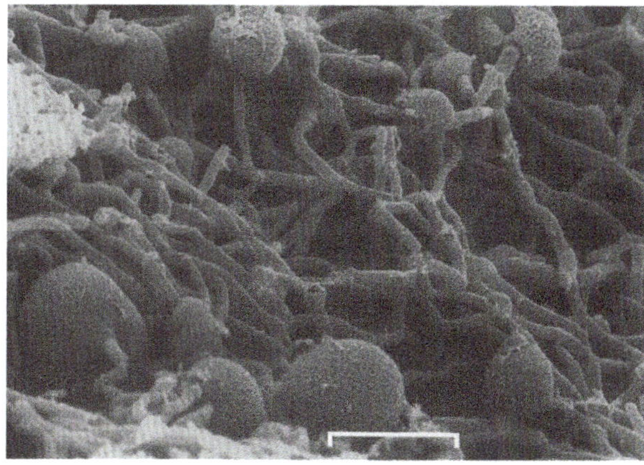

Fig. 6.16 *Pseudonacardia asaccharolytica*, bar = 5 μm. (Reichert et al. 1998)

Fig. 6.17 Biofilter of exhaust air cleaning of a belt dryer at the wastewater treatment plant in Balingen, Germany. (Heindl 2012)

Fig. 6.17 shows a biofilter for final cleaning of around 5000 m^3/h of exhaust air of a sludge belt dryer (Heindl 2012) and Fig. 6.18 the filling of a biofilter with wood chips (Heindl 2010).

High-end biofilters are covered with a roof and equipped with a water spraying device to keep the bed constantly wetted. In Fig. 6.19, a biofilter of this kind is covered with plastic housings and equipped with a chimney (Heindl 2022). The German *TA Luft* from 2021 stipulates that the odor of cleaned air leaving a biofilter is not allowed to resemble the odor of sludge. Typically, the basic odor of air from a biofilter that is used for sludge drying is around 200–300 odor units per m^3 (normalized to 20 °C [68 °F] and 101,325 Pa and in wet conditions) and is comparable to the odor of wet wood and moist soil.

The elimination rate of a biofilter for VOC is approximately linear proportional to the square root of the inlet VOC concentration. Fig. 6.20 demonstrates an example for the elimination rate of VOC expressed as g C (carbon) per m^3 of biofilter and hour (PCC Industries 2022). Assuming that exhaust air has a VOC load of 0.81 g/m^3 when entering the biofilter, the square root will be 0.9 and the specific elimination capacity is around 35 g $C/(m^3$ of biofilter × h).

Fig. 6.18 Filling of a biofilter with wood chips. (Heindl 2010)

Fig. 6.19 Covered biofilter for final odor treatment and chimney for releasing cleaned air into the environment. (Heindl 2022)

Fig. 6.20 Example of an elimination rate of volatile organic compounds (*VOC*) in a biofilter. (According to data from PCC Industries 2022)

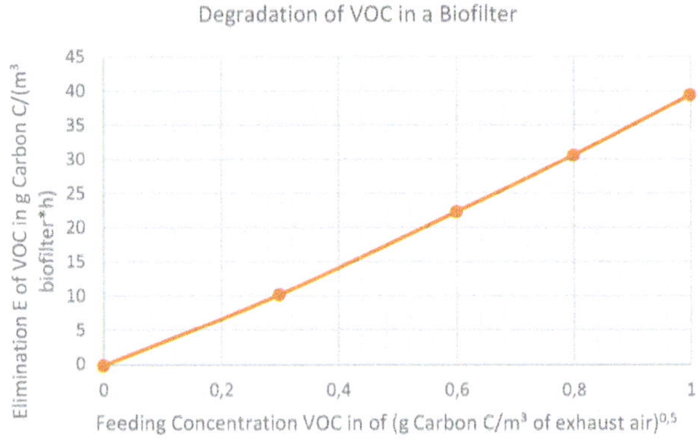

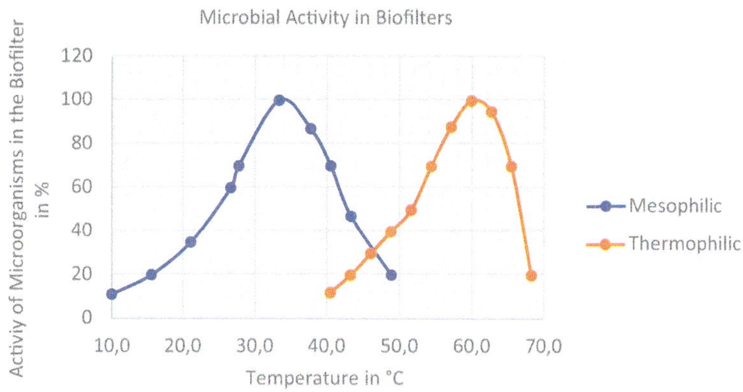

Fig. 6.21 Biological activity in biofilters with mesophilic and thermophilic operating modes. (According to data from PCC Industries 2022)

The approximate equation for the elimination capacity E in g C/(m^3 of biofilter × h) can be described with a linear function:

$$E = 39.68 \times C_{VOC}^{0.5} - 0.7674 \quad \text{with } R^2 = 0.9981 \tag{6.20}$$

with C_{VOC} in g/m^3 referring to the concentration of VOC in exhaust air entering a biofilter.

The microbial activity of mesophilic and thermophilic bacteria depends on the temperature of the saturated exhaust air entering the biofilter. Mesophilic operation of biofilters is common and temperatures range from a minimum of 10 °C (50 °F) to 49 °C (122 °F), with maximum activity at 33 °C (92 °F), as shown in Fig. 6.21. Thermophilic degradation of gas compounds is also possible when considering that the temperature of operation of biofilters ranges from 40 °C (105 °F) to 68 °C (155 °F). For thermophilic biofilters, maximum activity is reached at around 61 °C (Fig. 6.21 [PCC Industries 2022, modified]). Therefore, the biofilter temperature should always be kept in the temperature range of maximum activity to exploit the full cleaning capacity of the biofilter.

The lifetime of biofilters varies with filter material according to Table 6.17 (Störk 2022) and VOC load. The degradation of filter material is enhanced at pH-values below 6 and the lifetime reduced.

Biofilters are sometimes designed not only for the final treatment of dryer exhaust air but also to clean exhaust air from dewatering machines or small bunkers for the storage of

Table 6.17 Approximate lifetime of filter material used in biofilters

Type of biofilter material	Lifetime
Mixture of coconut fibers and fiber peat 50 : 50 to 70 : 30	Up to 6 years
Mixture of spruce wood chips and bark mulch 50 : 50, grain size 20–40 mm	Up to 4 years
Mixture of wood chips from pine, larch, Douglas fir, grain size 20–80 mm	Up to 5 years
Mixture of pine barkwood and pine wood, grain size 40–60 mm	Up to 5 years

dewatered sludge. Thus, the concentration of gas components decreases in some cases, but methane content, for example, may increase considerably. To conclude, there is a thinning effect that is accepted for exhaust air cleaning of wastewater treatment plants.

It is also possible to keep a biofilter in stand-by mode of operation by passing a smaller exhaust airflow through the biofilter bed and adding some nutrients to the water spraying device. In an emergency, the filter is able to process the full exhaust airflow within a shorter period far below the normal starting time for biofilters.

The VDI 3477 provides clear instructions for the proper operation of biofilters to maintain their functionality.

6.12 Activated Carbon Filters

In the case of a fresh air–exhaust air drying system, the exhaust air temperature is higher than 40–42 °C, which is the maximum allowable temperature for a mesophilic biofilter. Therefore, an adsorption filter bed of activated carbon is sometimes used instead of the mesophilic biofilter. Also, activated carbon filters are used as a final exhaust air purification stage.

Typical features of activated carbons and their application include:

- Exhaust air flow ranges from 0.1 to 0.5 m/s in the free-flow section of the adsorber (VDI 3674 [2013]), typically around 0.2 m/s.
- Thicknesses of activated carbon beds are between 500 and 1000 mm, in some cases up to 1500 mm, depending on the use of vertical or horizontal bed filters.
- Steam activated or chemically impregnated carbons are used, depending on where their main task lies: preferably physical or chemical adsorption.
- The best adsorption capacity is achieved the lower the temperature of exhaust air/exhaust gas is adjusted.
- With a few exceptions, lower relative gas humidities below 60% favor adsorption of gas molecules.
- Staged, serial arrangement with improved exploitation of adsorption capacity.
- Ability to remove siloxanes.
- Final cleaning stage to remove residual amounts of gas components.

In any event—as activated carbon is expensive—these filters should be used only for final cleaning of exhaust air streams to reduce odor and/or VOC load below limits set out in local regulations. Loading activated carbon with a high mass flow of VOC cannot be carried out on an economic basis.

The carbon is sometimes impregnated with NaOH, Na_2CO_3, K_2CO_3, KJ, or KOH for improved separation of sulfuric components that have not been completely removed by the alkaline scrubber. Thus, an alkaline surface in the inner capillary is created to promote chemical adsorption at pH-values normally higher than 7.5. The alkaline impregnation is around 2–5 mass% of the weight of the dry activated carbon. For example, methane thiol (CH_4S) reacts with the alkaline substances under formation of water or carbonic acid:

$$CH_3SH + NaOH \Leftrightarrow CH_3SNa + H_2O \quad (6.21)$$

$$2\,CH_3SH + K_2CO_3 \Leftrightarrow 2\,CH_3SK + H_2CO_3 \quad (6.22)$$

As a rule of thumb, around 100–200 mg of sulfur can be loaded per g of activated carbon in the case of pure sulfuric components such as H_2S. However, this value strongly depends on the type of sulfuric gas component and additional reactions of further gas components at the inner surface of activated carbon, displacement of smaller VOCs by larger VOCs, as well as on the concentration of the gas component in exhaust air.

Before the exhaust air enters the activated carbon filter, it needs to be heated at least 5–7 K above its dew point temperature to avoid any condensation in the filter system. Condensation of water vapor in small capillaries active in the adsorption of odorous gas substances must be avoided. Fig. 6.22 shows the dependence of the relation of steam pressure to saturated steam pressure—corresponding to relative humidity of air—on the temperature and radius r of capillaries, calculated by Thomson's formula. At a temperature of operation of 50 °C, the relative humidity—corresponding to the relation of steam pressure to saturated steam pressure—must be less than 81% to avoid condensation in capillaries with a radius of less than 2 nm and less than 66% with a pore radius of less than 1 nm. In any event, insulation of the activated carbon filter is recommended to avoid condensation at cooler walls of the filter thus reducing the efficiency of adsorption in filter bed by condensate diffusion.

The affinity of the inner surface of activated carbon to volatile substances is high in the case of volatile carbon acids, alcohols, and esters. It increases with augmenting polarity of the adsorbent (Bathen and Breitbach 2001). For adsorption of VOCs in general, water vapor-activated carbons are used with a high inner surface and a pore distribution, allowing larger VOCs to penetrate into the inner surface of the granules.

The small micro-pores of 1–3 nm in diameter—or 0.5–1.5 nm in radius—have the highest adsorption potential and determine the adsorption quality of an activated carbon. The meso-pores with a diameter of around 8–40 nm mainly act as multi-layer adsorbers, and in some cases condensation in capillaries occurs. The macro-pores serve as transport channels for VOCs to be separated (Timofejew 1967).

Adsorption of VOCs on activated carbon increases with augmenting concentration of VOC and decreases with higher gas temperatures. For the gas component toluene (C_7H_8), a hydrocarbon occurring in exhaust air of sludge dryers, the maximum absorption at a concentration in exhaust air of 100 mg C_7H_8 as a single gas component is around 17 mass% at 20 °C, but declines to around 7.5% at 70 °C on a steam-activated carbon based on hard

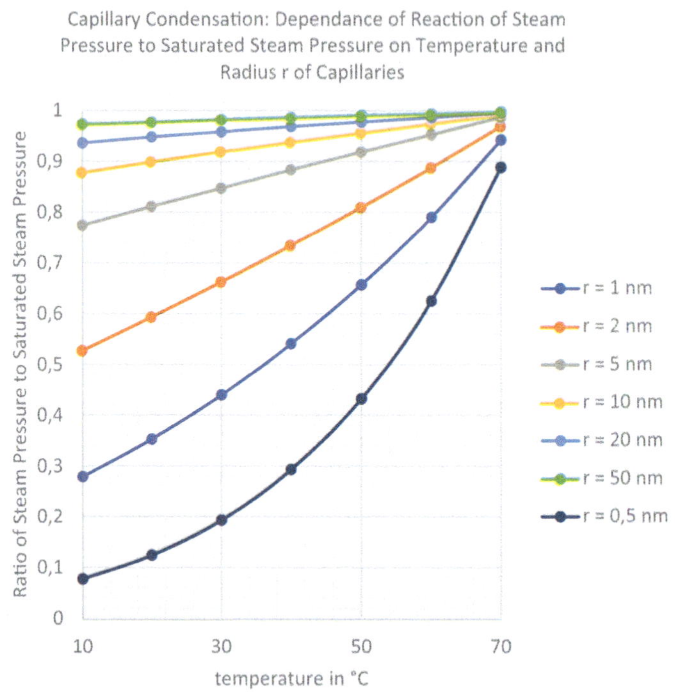

Fig. 6.22 Capillary condensation depending on temperature and capillary radius

coal (Donau Carbon 2020). It is not possible to predict the adsorption capacity of single components in a mixture of a myriad of different gas components. Single gas adsorption is usually lower in an exhaust gas containing different gas components.

The maximum adsorption in g gas component per g of an activated carbon corresponding to a balance condition decreases with increasing temperature and increases with augmenting concentration of the gas component in the exhaust air. Treatment of gas mixtures always results in lower adsorption rates of individual gas components, which will be smaller than with single gas components.

Table 6.18 shows typical qualities of an activated carbon used for exhaust air treatment to remove sulfuric components (Donau Carbon 2020), valid for single gas components.

A further alkaline-impregnated activated carbon is DESOREX G 50, which is designed for adsorption of H_2S, being chemically bound. This coal is also suitable for binding organic substances. Its bulk density is $600 \pm 60\,\mathrm{kg/m^3}$. The cylindrically shaped particles with a diameter of around 4 mm show a total Brunauer–Emmett–Teller (BET) inner sur-

Table 6.18 Typical properties of an activated carbon used for the removal of sulfuric components. (Donau Carbon 2020)

Type of activated carbon	Donau carbon Desorex PI 50 K or PI 53 K
Type of reaction on surface of activated carbon	Catalytic–chemical
Raw material	Bituminous coal
Condition for operation (PI 50 K)	Oxygen in exhaust air
Brunauer–Emmett–Teller (BET) surface	700–1100 m^2/g
Bulk density	$500 \pm 30\ kg/m^3$
Particle size	≈ 4 mm
For removal of:	Hydrogen sulfide (H_2S), methane thiol (CH_4S), methylamine (CH_3NH_2), ammonia (NH_3), indole (C_8H_7N), skatole (C_9H_9N)
Limits of operation	Temperature: 10–70 °C Relative humidity: < 70–95%
Iodine adsorption (PI 50 K)	1000 ± 50 mg/g of activated carbon
Adsorption of CCl4 (PI 50 K)	Maximum 60 mass%
Maximum H_2S concentration at inlet of filter	100 ppm, higher concentrations are harmful to the activated carbon
H_2S breakthrough concentration (PI 50 K)	$> 0.12\ g/cm^3$ of activated carbon
Minimum H_2S saturation	0.2 g/g of activated carbon
Water content relative to total mass (PI 50 K)	< 5% when packed
Pressure loss for 4 mm activated carbon granules at 20 °C and 100,000 Pa	0.1 m/s linear exhaust gas speed: ≈ 300 Pa/m 0.2 m/s: ≈ 800 Pa/m
Porosity of filter bed	0.3–0.4

face area of around 950 m^2/g. The weight of impregnation (NaOH) is around 5 mass%, the CCl_4 exceeds 50 mass%. G 50 activated carbon shows low sensitivity to high relative gas humidities (Donau Carbon 2003).

Carbon tetrachloride (CCl_4) activity expressed in adsorbed mass% of the weight of activated carbon is a measure for the pore volume according to ASTM 3467. Nowadays, the butane (C_4H_{10}) activity is applied according to ASTM D5742 and recalculation can be performed by:

$$\text{Mass\% } CCl_4 = 2.55 \times \text{mass\% } C_4H_{10} \quad \text{for } \frac{p}{p_s} = 0.9 \tag{6.23}$$

H_2S extraction increases with augmenting temperature and relative humidity of exhaust air. The hydrogen sulfide is transformed to elementary sulfur in the pores of the activated carbon, which acts as a catalyst:

$$2\,H_2S + O_2 \Leftrightarrow 2\,S + 2\,H_2O \tag{6.24}$$

A further reaction of elementary sulfur to sulfur dioxide is possible:

$$S + O_2 \Leftrightarrow SO_2$$

At temperatures exceeding 70 °C, a side reaction begins on the surface of activated carbon resulting in the formation of sulfuric acid diffusing into exhaust air:

$$H_2S + 2\,O_2 \Leftrightarrow H_2SO_4 \tag{6.25}$$

In this case, the downstream components of the exhaust air treatment system are at risk of corrosion.

The maximum load of methane thiol on bituminous coal is reported to be more than 200 mg/g of activated carbon (Bashova et al. 2003). However, all the data on maximum load are valid *for single gas components*, whereas the exhaust gas of a sewage sludge dryer contains a mixture of several hundred gas components with different concentrations. Some of these components may compete during adsorption on the surface of the activated carbon, resulting in desorption of already adsorbed components and thus decreasing the adsorption capacity for a certain gas component.

In addition, small gas components that are only adsorbed physically at the inner surface of the activated carbon are displaced by larger ones due to lower adsorption energy and pushed through the bed of activated carbon. If these small components have low odor thresholds or high osmogenous potentials, they dramatically will determine the lifetime of the activated carbon filter. Therefore, far lower loads of VOC mixtures below 100–200 mg VOC/g of activated carbon are occasionally observed when odor load components dominate the lifetime of an activated carbon filter. On the other hand, the adsorption capacity of activated carbon is also considerably reduced with increasing temperature and it decreases with falling concentration of gas components in exhaust air. In extreme cases, only 10 mg/g can be achieved, making economic operation virtually impossible.

Minimum linear contact times in activated carbon beds for different adsorption phenomena are listed in Table 6.19. Prolonged contact times far beyond the indicated values may have a negative impact on the lifetime of activated carbon as smaller gas components that are physically adsorbed will be displaced by compounds of higher molecular mass, thus limiting the lifetime of the filter.

In practical filtering, the mass transfer zone—separating the fully saturated area of activated carbon bed from the non-charged bed zone—has an 'S' shape whose downstream peak determines breakthrough either due to odor units or due to VOCs. Therefore, the width of the mass transfer zone should be kept as small as possible. This can be achieved by (von Kienle and Bäder 1980):

- Increasing the bulk density of activated carbon
- Decreasing linear gas velocity in the bed
- Keeping the diameter of activated carbon granulate small
- Reducing the gas temperature in the bed
- Operating at higher VOC concentrations

In addition, a broad range of VOCs from small to very large also increases the width of the mass transfer zone, and the gas components with carbon contents of C = 1–5 will be

Table 6.19 Minimum linear contact time for activated carbons

Type of activated carbon	Minimum linear contact time in seconds
Impregnated carbon with chemical sorption	≥ 2
Water-activated carbon with physical adsorption	≥ 1–2
Catalytic reactions after UV-C treatment of exhaust air	≥ 2–3

the first to be discharged from the carbon bed (Bratzler 1944). Trials with dry air charged with tetrachlorethylene (C_2Cl_4) ($12\,\mathrm{g/m^3}$) at ambient temperature showed that the width of the mass transfer zone in an activated carbon bed ranged from 2.5 to 16 cm, depending on carbon granulate diameter and linear gas velocity (von Kienle and Bäder 1980). For a 4-mm granulate diameter d, the width b of the mass transfer zone in cm could be described for this trial by:

$$b = 100 \times w_{\mathrm{gas}}^2 + 80 \times w_{\mathrm{gas}} + 1 \tag{6.26}$$

with w_{gas} as linear gas velocity in m/s. With a gas velocity of 0.2 m/s, the formula for the width b of the mass transfer zone in cm is calculated by:

$$b = 1.2712 \times d^{1.6814} \tag{6.27}$$

with d being the granulate diameter of the activated carbon (von Kienle and Bäder 1980).

The activated carbon can be delivered in special mobile exchange containers, which go into a reactivation process when carbon is saturated. A container with reactivated carbon is installed in the position of the replaced container. Reactivation is normally carried out only with water vapor-activated carbon and not with impregnated carbon. In any event, the sulfur content of loaded activated carbon must be less than 2 mass%.

The breakthrough or saturation point of an activated carbon filter bed is determined during the first period of operation, so that the replacement in containers is organized in time. Another option is to choose a key component, usually an odorous gas component of low molar mass, which is liable to be replaced by larger gas components.

Very often, odor limits are exceeded before the full adsorption capacity for VOCs of the activated carbon is exploited. Therefore, a staged inline arrangement of two filters with optionally an additional change of sequence as soon as one filter is filled helps to considerably improve exploitation of filter capacity.

6.13 Boosting of Exhaust Air Treatment in the Drying of Sludge

As already mentioned, exhaust air treatment creates greater requirements in the case of drying of sludge from external sources that has been mechanically stressed, stored for a longer period, and exposed to intensive oxygen access. As a consequence, far higher loads of VOCs may be encountered in exhaust air and additional equipment is needed. The exhaust air treatment system can be equipped with:

- Acid scrubber + caustic scrubber + biofilter (standard for drying of sludge produced at wastewater treatment plants and the use of a condensing dryer). VOC removal in a caustic scrubber can be improved by mixing ozone (O_3) (at higher concentrations of $> 130\,g\ O_3/Nm^3$ of ozone/air mixture) or an organic solvent such as polyethylene glycol or silicone oil into the recirculation water piping of the scrubber. The latter method improves mass transfer from the gas phase into the liquid phase of the scrubber, but costs of operation increase.
- Acid scrubber + UV-unit (ultraviolet radiation with a wavelength of 185 and 254 nm) + activated carbon filter acting as a catalyst (for higher contents of VOC and gas components with low water solubility).
- Acid scrubber + RTO (regenerative thermal oxidation; for very high contents of VOC and gas components with poor or no water solubility). However, this means additional consumption of natural gas to maintain the necessary temperature in the oxidation process (roughly 1 m^3 of natural gas for 1000 m^3 of exhaust air, depending on VOC content of exhaust air and calorific value). More economical use can be achieved by adding exhaust air from the storage of dewatered sludge due to its high methane content.

Owing to the high ammonia content in dryer exhaust air, the use of an acid scrubber is envisaged due to its high efficiency of NH_3 elimination. However, the minimum concentration of NH_3 after treatment is limited to approximately 1 mg/m^3 under full-scale operational conditions.

Attention must be paid to the content of volatile silicone compounds in the raw exhaust gas leaving the dryer, as formation of SiO_2 deposits will cause more frequent cleaning operations in the case of UV-units and RTO units. Usually, in the case of stored sludge and in belt drying of sludge, up to 2 mass% of the VOC charge in exhaust air is silicon (Si).

The catalytic action of activated carbon with regard to oxidation of H_2S depends on the amount of oxygen surface complexes and the proportion of pores 3.5–8 nm in diameter (Siedlewski 1961).

6.13.1 Regenerative Thermal Oxidation

When using a regenerative thermal oxidation (RTO), one has to bear in mind that VOCs are transformed to NO_x and SO_2. These gas components produce an odor impression of incineration, i.e., the odor threshold of NO_2 at 20 °C is approximately 0.3548 mg/m^3 and that of SO_2 is around 1.8621 mg/m^3 (Devos et al. 1990). The evaluation of this odor impression depends on local authorities, in contrast to German *TA-Luft* regulations for sludge drying, odor units of up to 3000 per m^3 (himid condition, 20 °C, 101,325 Pa) have been accepted in the past for RTO exhaust gas. The retention time of exhaust air in the RTO is around 1 s at 800–820 °C (1498–1534 °F). A further literature source stipulates a range of 0.5 to 1.5 s at temperature levels of 704 to 980 °C (WEF 2020). In the presence of NH_3 in higher concentrations, the oxidation temperature is around 850–875 °C ($\approx$ 1560–1600 °F), but there is a heterogeneous catalytic effect on the surface of the hot porous ceramic stones enhancing gas reactions. RTOs have a very high thermal efficiency of up to 95% due to their regenerative heat recovery in chambers of perforated ceramic stone blocks or ceramic fillers that are flown through by turns, a further sources indicates even 99% of efficiency (WEF 2020). In the case of gas humidity approaching 95–100%, preheating of exhaust gas of 10–15 K is necessary to avoid condensate formation and odor carry-over by re-evaporation during change of flow through of heat recovery chambers. Superficial velocities in the holes of the ceramic blocks typically range from 1.3 to 1.4 m/s.

NH_3 is transformed to NO by a mass ratio of around 15–40% according to practical experience:

$$2\,NH_3 + 2.5\,O_2 \Leftrightarrow 2\,NO + 3\,H_2O \tag{6.28}$$

At temperatures below 700 °C, formation of NO_2 starts:

$$2\,NO + O_2 \Leftrightarrow 2\,NO_2 \tag{6.29}$$

Another portion of NH_3 reacts to N_2 and H_2O and the residual mass of ammonia slips through. This reaction is accelerated with increasing temperature and by surface catalysis.

$$4\,NH_3 + 3\,O_2 \Leftrightarrow 2\,N_2 + 6\,H_2O \tag{6.30}$$

One must consider that the chemical balance of these gas reactions is not completely achieved.

For practical operations, one needs to consider (personal communication of Dürr 2022) that:

- Scavenging fresh and heated air must be used for cleaning purposes with a maximum mass portion of 5–10% over a time portion of 20% of total operation time.
- A gas burner needs to be operated to compensate for heat losses in the range of 1.0–1.2 m^3 of natural gas/h per 1000 m^3 of exhaust air, as a self-sustaining operation on the basis of calorific value if VOC up to 200 mg C/m^3 (at operating temperature) cannot be achieved. It can be estimated that at least 1200–1500 mg C/m^3 of exhaust air is necessary to avoid additional gas heating in steady-state operation. The calorific situation can be improved by adding exhaust air from the bunker for dewatered sludge with a comparatively high content of methane to dryer exhaust air in front of the RTO.
- A high content of siloxanes in exhaust air will increase deposits of SiO_2 and increase cleaning as well as maintenance requirements.

For thermal treatment of exhaust air, the German *TA Luft* (2021) stipulates limits for carbon monoxide (CO), sulfur dioxide (SO_2), and nitrogen oxide, expressed as NO_2 (Table 6.20) (TA Luft 2021).

Usually, the carbon monoxide and sulfur dioxide limits can be adhered to without problems, but depending on NH_3 content of exhaust air, there will be a higher concentration of NO_x in flue gas. Therefore, the use of an acid scrubber in front of the RTO is rec-

Table 6.20 Limits for exhaust air treatment by regenerative thermal oxidation (RTO). (According to the German *TA Luft* 2021)

Gas components in exhaust gas of RTO	Maximum concentration in mg/Nm^3_{dry}
NO_2	100
CO	100
SO_2	350

ommended. According to practical experience, siloxane contents need to be far less than 5–10 mg siloxanes per m^3 of exhaust air in front of an RTO unit to avoid blocking of heat transfer stones by SiO_2 for an acceptable period of operation.

6.13.2 Ultraviolet Radiation

Ultraviolet (UV) radiation applied to exhaust air of belt dryer acts in three ways:

- Generation of ozone (O_3) from oxygen (O_2) and hydroxyl (OH)· radicals from water vapor in exhaust air, caused by radiation at a wavelength of 185 nm (typical for low pressure mercury or amalgam UV lamps) and reaction of these oxydising agents with VOCs.
- Rupture of bonds of gas components by UV photons of high energy density, caused by radiation at a wavelength of 254 nm (typical for low pressure mercury and amalgam UV lamps).

It is of utmost the importance that the selection of UV lamps and the total electric power for a given application is related to the overall VOC throughput carried by exhaust air. In addition, the surface temperature of lamps appearing in operation and extending from the feeding to the discharging end of the UV reactor must be as close as possible to the temperature of maximum UV-radiation output of lamps at 185 nm. In this context, the speed of exhaust air around the lamps acts as a cooling agent and influences lamp surface temperature. A flow and heat transfer simulation helps in gauging temperatures of lamps. Trials with test units securing transferability with reasonable upscaling factors are strongly recommended.

The use of UV-units for exhaust air treatment is based on:

- Exhaust airflow in free space between horizontally arranged UV lamps: $\approx$2–6 m/s, air flow cooling of lamps is important to keep surface lamp temperature as far as possible close to the optimum surface temperature range with the highest UV radiation efficiency.
- The distance of lamps is $\geq$ 100–200 mm for thermal heating reasons; as a rule, lamps are mounted in a staggered arrangement.

- Amalgam lamps, low-pressure type, covering wavelengths of 185 and 254 nm with a maximum efficiency at lamp temperatures of around 95–105 °C (Heraeus 2021).
- Mercury lamps, low-pressure type, covering wavelengths of 185 and 254 nm with maximum efficiency at lamp temperatures of around 38–45 °C (see Fig. 6.23; Heraeus 2021).
- Typical UV radiation efficiencies at 254 nm wavelength for low-pressure mercury lamps at the optimum lamp temperature of around 40 °C range from 31 to 34%—relative to electric energy consumption—and for amalgam lamps also at the optimum lamp temperature of 95–105 °C; for the wavelength of 185 nm, UV radiation efficiency ranges from 6 to 9%—relative to electric energy supplied to lamps—for both low-pressure lamps with synthetic quartz glass.
- For low-pressure mercury lamps at lamp temperatures exceeding the optimum temperature, the decline in UV radiation efficiency for the 254 nm wavelength can be estimated to be around 14–20% relative per 10 K of temperature increase; for the 185 nm wavelength, the reduction can reach around 5–7% relative. Of course, these reduction values depend on lamp diameter as well as lamp design and can vary considerably.
- For amalgam lamps surface temperatures for optimum process applications with a broad temperature area of more than 90% of UV-efficiency can be adjusted by special alloys added to the mercury, e.g. amalgam mixtures like Bi58Sn38In3Hg0.7 (Bi: bismuth, Sn: tin, In: Indium, Hg: mercury with indication of mass-%) (Elektrische Glühlampen mbH 2008).
- Synthetic quartz glass for long-life operation with ongoing transparency at the 185 nm wavelength. At wavelengths equal to and below 185 nm, tridymite, an opal-white variety of quartz, is formed in the quartz matrix and decreases transparency of quartz glass to UV radiation, thus limiting lifetime (Summer 1962). Only with high-quality synthetic quartz glass, a maximum efficiency of 9% at 185 nm can be achieved.
- Inner coating made of aluminum or yttrium oxides as a protective layer against HgO formation of layers absorbing UV radiation (Van der Meer et al. 2023) to prolong lamp lifetime.
- Lifetime of UV lamps of approximately 10,000–12,000 h for load change on/off of less than 500/a, with a reduction of UV radiation to 80–85% of the initial value.
- Regular CIP cleaning of lamps necessary—warm water with 50–70 °C and a hardness of less than 5°dH and almost free of silicate—for removal of silicon oxide (SiO_2) deposits on the lamp surface. Alkaline cleaning agents such as sodium hydroxide (NaOH), potassium hydroxide (KOH), sodium carbonate (Na_2CO_3), sodium phosphate (Na_3P), and alkali-based soaps as well as sodium hydrogen sulfate ($NaHSO_4$) are not recommended for cleaning purposes, as they tend to attack the quartz glass, forming a milky layer that reduces UV radiation. Tripolyphosphate ($H_5P_3O_{10}$), nitric acid (HNO_3), formic acid (CH_2O_2), acetic acid ($C_2H_4O_2$), citric acid ($C_6H_8O_7$), amidosulfonic acid (H_3SO_3N), uric acid ($C_5H_4N_4O_3$), phosphoric acid (H_3PO_4), hydrogen peroxide (H_2O_2), aldehydes ($C_xH_yC{=}O$), alcohol ($C_xH_y{-}OH$), and peracetic acid ($C_2H_4O_3$) are suitable for CIP cleaning (Heraeus personal communication 2023).

- Minimum linear gas retention time of 2–3 s in activated carbon, where O_3 takes over the final oxidation of VOCs on the inner surface of water vapor-activated carbon. However, this catalytic effect only occurs as long as ozone molecules are able to adsorb on a free inner surface area of activated carbon and react with a VOC molecule nearby.
- Activated carbon is made of black coal or coconut shell, activated by water vapor without chemical impregnation; typical data for an activated carbon on the basis of coconut shell are given in Table 6.21 and pore distribution in Fig. 6.23.
- Limitation of lifetime of activated carbon by deposits of sulfur substances and VOC fragments and amount of available ozone for catalytic decomposition of VOC fragments at the inner surface of activated carbon.
- The design of electric power of UV lamps is related to the mass flow of VOC in Wh/g VOC; the normal typical design for odor reduction only is carried out to 1000 m^3 of exhaust air and covers 1.0–2.0 kWh/1000 m^3 and is not applied in the case of VOC and odor removal of exhaust air of belt dryers for sewage sludge at high VOC loads. Actually the UV-radiation in the UV-reactor itsself results in a production of smaller VOCs and if occuring only in a partial oxidation of VOCs. The VDI 2441, however, indicates a range of 0.3–4.5 kWh of electric energy/1000 m^3 of exhaust air for odor control systems; for VOC control systems, the stipulated values cover an even broader range of values of 0.5–12 kWh/1000 m^3/h (VDI 2441 2016). It must be mentioned

Table 6.21 Typical data of a water vapor-activated carbon used for final catalytic oxidation after UV-C exhaust air treatment

Type of activated carbon	Donau carbon UC-50
Basis	Coconut shell
Treatment	Water vapor activation
Brunauer–Emmett–Teller (BET) surface	$\approx 1100\,m^2/g$
Total pore volume	$\approx 0.77\,cm^3/g$
Specific micro-pore volume (radius 0.2–1.0 nm)	$0.444\,cm^3/g$
Specific meso-pore volume (radius 1.0–25.0 nm)	$0.071\,cm^3/g$
Specific macro-pore volume (radius > 250 nm)	$0.255\,cm^3/g$
Grain size	4×8 mesh (2.36×4.75 mm)
Bulk density	$480 \pm 50\,kg/m^3$
Water content when packed	< 5 mass%
Ash content when packed	< 5 mass%
CCl_4 adsorption	> 50 mass%
Iodine adsorption	1000 mg/g
Benzene adsorption from air at 20 °C at $p/p_s = 0.9$ (saturation at 20 °C 320 g/m^3)	32 mass%
Ball pen hardness	> 95 mass%
Air flow resistance (20 °C, 1 bar[a])	0.1 m/s: $\approx$ 270 Pa/m 0.2 m/s: $\approx$ 820 Pa/m 0.3 m/s: $\approx$ 1600 Pa/m

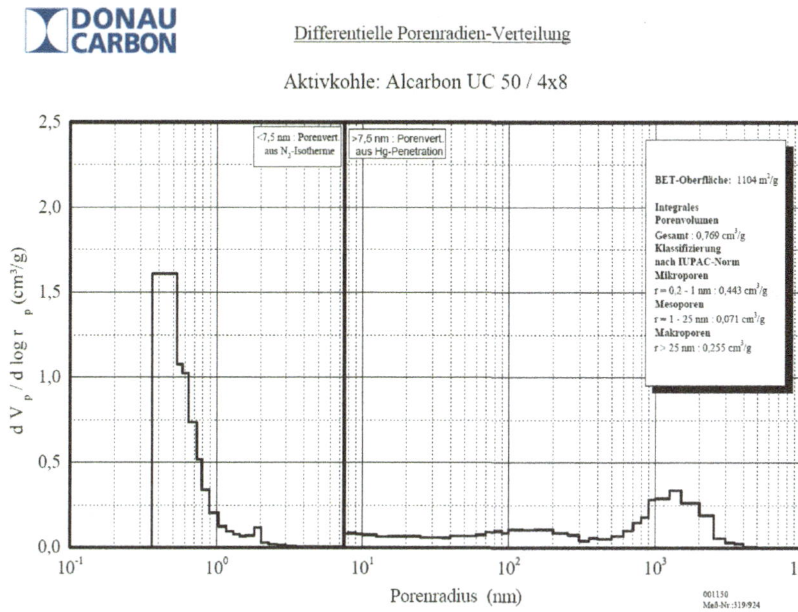

Fig. 6.23 Pore distribution of water vapor activated carbon UC 50. (Donau Carbon 2022). *Aktivkohle*, acitvated carbon; *Differentielle Porenradienverteilung*, differential pore radius distribution; *BET-Oberfläche*, BET surface; *Integrales Porenvolumen*, integral pore volume; *Mikroporen*, micropores; *Mesoporen*, mesopores; *Makroporen*, macropores; *Integrales Porenvolumen*, integral pore volume; *Klassifizierung nach IUAPC-Norm*, Classification according to IUAPC standard; *IUAP*, International Union of Pure and Applied Chemistry

that the real electric energy demand strongly depends on the UV lamp efficiencies determined by the lamp surface temperature and the type of odor and VOC to be treated.

- Methane (CH_4) cannot be destroyed on an economical basis with normal low-pressure lamps (destruction $\ll$ 10 mass%).
- Usually, 316L-grade stainless steel is used for UV units due to corrosion risk, as SO_2 and SO_3 are formed by the oxidation of sulfur components in exhaust air, so that sulfuric acid is formed by a reaction of SO_3 with condensate (low pH-value of typically $\approx$ 2.5 to 3.5) of exhaust air.
- Airflow resistance of UV units ranges from approximately 400 to 800 Pa, depending on reactor length and number of lamps, air speed, lamp arrangement, airflow control flaps, demisters, and separation of individual sections with perforated plates or air guiding elements near to lamps.
- Efficiency of UV-radiation and further chemical reactions strongly depends on hit probability of radiation and proper mixing of reacting gas components.

UV ozone generation from oxygen of exhaust air is related to UV radiation at 185 nm. For low-pressure mercury lamps, up to around 54 g of O_3 are formed per kWh of electric energy supplied to UV radiation of 185 nm (Voronov 2008). Low-pressure mercury lamps with a 9% energy portion of radiation at 185 nm produce around 48.6 g ozone per kWh of electric energy transferred to the UV lamp. However, the radiation of 254 nm destroys a portion of the produced ozone, with the degree depending on the O_3 concentration in surrounding gas. Therefore, as a rule of thumb, it is reported that around 6–12 g ozone per kWh effectively remains (Claus 2021). Voronov measured an ozone generation rate of around 17.5 g/kWh of total electric energy at an ozone concentration of around 17 mg/m^3 of air, giving a destruction rate of ozone by 254 nm radiation of approximately 64% (Voronov 2008). The data are valid for lamp surface temperatures around the optimum point of 40 °C.

The three main reactions of UV radiation with exhaust gas can be described by the formulas (VDI 2441 2016):

Photolysis/cracking of VOCs:

$$\mathrm{R{-}R} + \mathrm{UV{-}C} \Rightarrow \mathrm{R\cdot} + \mathrm{R\cdot} \tag{6.31}$$

with R· as cracked/radical components.

Ozone formation:

$$\mathrm{O_2} + \mathrm{UV{-}C\,(185\,nm)} + \mathrm{M} \Rightarrow 2\,\mathrm{O\cdot} + \mathrm{M} \tag{6.32}$$

$$\mathrm{O_2} + \mathrm{O\cdot} + \mathrm{M} \Rightarrow \mathrm{O_3} + \mathrm{M} \tag{6.33}$$

OH radical formation:

$$\mathrm{H_2O} + \mathrm{O\cdot} \Rightarrow 2\,\mathrm{OH.} \tag{6.34}$$

$$\mathrm{H_2O} + \mathrm{UV} - \mathrm{C} \Rightarrow \mathrm{OH\cdot} + \mathrm{H\cdot} \tag{6.35}$$

The · indicates a radical condition of the gas component.

The ends of the curves in Fig. 6.24 are elongated by the author to higher temperatures for mercury lamps and to lower temperatures for Amalgam lamps—the optimum temperature range for mercury lamps is ≈ 40–45 °C, for amalgam lamps typically ≈ 95–105 °C. It must be emphasized that the absolute values of UV efficiencies and their decline, besides optimum lamp surface temperatures, depend on lamp construction, quartz glass diameter, and pressure of mercury/amalgam/inert gas (aargon/neon) mixtures (Van der Meer et al. 2023). Therefore, there are also amalgam lamps that have their optimum of UV generation at lamp surface temperatures of 55 or 112 °C (Heraeus Noblelight 2023). The lamp surface temperature itself is influenced by the heat generation of the lamp itself (≈ 110 W electric power supplied/m of length for mercury lamps and ≈ 210 W/m for amalgam lamps) and by the cooling effect of the exhaust air flow (temperature of exhaust air, air speed, heating of exhaust air when passing through the UV reactor).

Fig. 6.25 demonstrates the energy of photons depending on the wavelength of UV radiation.

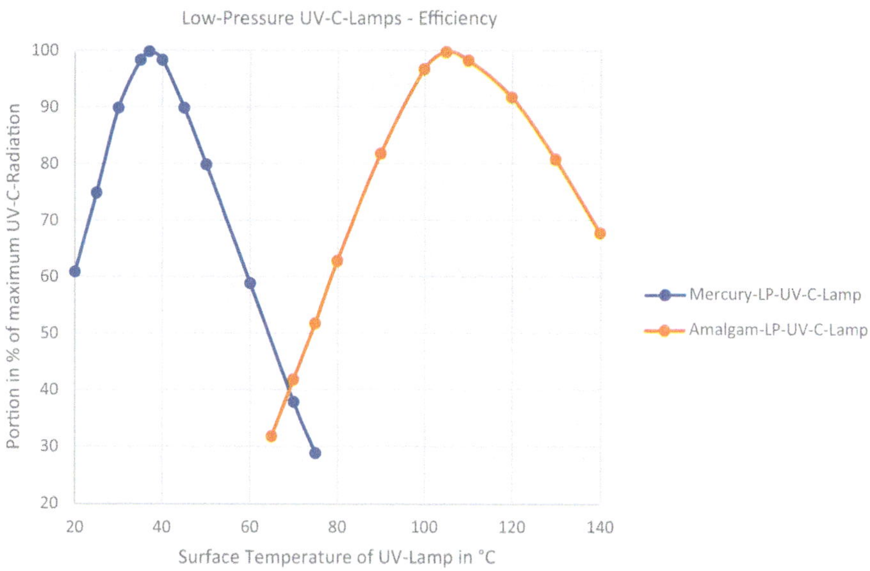

Fig. 6.24 Influence of lamp surface temperature of low pressure UV lamps on their efficiency with prolongation of curves to higher temperatures in case of mercury lamps and to lower temperatures in the case of amalgam lamps by the author. (Heraeus Noblelight 2021)

The photon energy W in eV is related to the wavelength λ in nm in formula (6.36).

$$W = 2 \times 10^{-9} \times \lambda^4 - 3 \times 10^{-6} \times \lambda^3 + 0.0013 \times \lambda^2 - 0.2874 \times \lambda + 30.887 \quad (6.36)$$

with a coefficient of determination $R^2 = 0.9997$.

Photon energy can also be calculated as follows:

$$W = \frac{1,239.6}{\lambda} \quad (6.37)$$

with λ in nm and W in eV (Summer 1962).

A low-pressure UV lamp has two main wavelengths: one at 185 nm and one at 254 nm. The energy of emitted photons is able to rupture most of the bonds of gas components, as described in Table 6.22. The energy values for the bonds without the indicated chemical substance are approximate and vary depending of chemical structure of substance.

The wavelength of 185 nm is the most effective. However, C≡C and C=O bonds cannot be ruptured by photons with a wavelength of 185 or 254 nm. The wavelength of maximum absorption of ammonia (NH_3) is close to 185 nm; therefore, it is important to remove ammonia in an acid scrubber in front of the UV unit as far as possible. Thus, exploitation of 185 nm for VOC treatment is improved. Nitrogen absorbs UV radiation in the wavelength range of less than 175 nm and greater than 300 nm. Therefore, there is no absorption at the typical UV wavelengths of 185 and 254 nm of mercury or amalgam

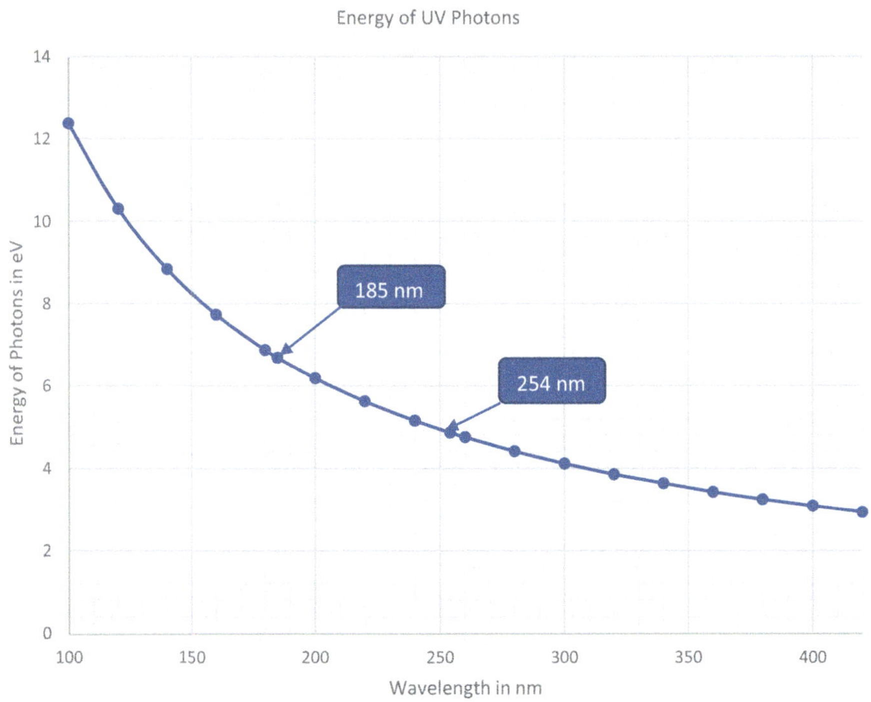

Fig. 6.25 Energy of photons depending on the wavelength of UV radiation; 185 and 254 nm represent the predominating wavelength of low-pressure mercury and amalgam lamps

low-pressure UV lamps. Table 6.22 also demonstrates that there are gas components with higher binding energy–or ionization energy–than the photon energy of the wavelength of 254 and 185 nm and that photooxidation of these gas components is not possible from the theoretical point of view.

Dimethyl sulfide is transformed under UV radiation of a wavelength of 254 nm (Dürr 1975):

$$(CH_3)_2S + UV(254\,nm) \Rightarrow C_2H_6 + (CH_3)_2S_2 + CH_4S + CH_4 \tag{6.38}$$

The mass portions of the generated compounds are 41% C_2H_6, 29% $(CH_3)_2S_2$, 11% CH_4S, and 3% CH_4. Of course, generated chemical compounds besides methane show further reactions with photons. For example, CH_4S is transformed to (Dürr 1975):

$$CH_4S + UV - C(254\,nm) \Rightarrow (CH_3)_2S_2 + H_2 + H_2S + CH_4 \tag{6.39}$$

Due to the low binding energy of the S–S bond, dimethyl disulfide is easily cleaved to CH_3S radicals at a wavelength of 254 m.

The electrical discharge in the radiation field is executed within 10 ns. The first chemical reactions follow a time range of micro- to milliseconds. Diffusion, heat, and mass

Table 6.22 Energy required for breaking up bonds in eV. (Semjonow 1961; Ellis and Wells 1941; Fehnel and Carmack 1949; Heraeus Noblelight 2021; Summer 1970; Franke 2011; NIST)

Gas component and bond	Energy required to break up the molecule or individual bond in eV	UV wavelength with maximum absorption or corresponding wavelength for breaking up of the bond in nm
Water vapor (H_2O)		≈ 184
Ammonia (NH_3): N–H	4.43	≈ 192
Acetone (C_6H_6O)	9.70	–
Acetic acid ($C_2H_4O_2$)	10.65	–
Benzene (C_6H_6)	–	< 200
Hydrogen sulfide (H_2S): H–S	4.00	≈ 185
Methane thiol (CH_4S): C–S	3.04	≈ 195
Dimethyl sulfide [$(CH_3)_2S$]: C–S	3.17	≈ 211
Dimethyl disulfide [$(CH_3)_2S_2$]: S–S	3.10	≈ 252
Sulfur dioxide (SO_2)	–	≈ 290
Methane (CH_4): C–H	4.40	≈ 122
Ozone (O_3)	–	253.9
Nitrogen (N_2): N≡N	9.76	Corresponding to ≈ 127 nm
Nitrogen oxide (NO)	–	≈ 226
Di-nitrogen oxide (N_2O)	–	≈ 185
Individual bonds	–	–
C–C	≈ 3.6	≈ 363
C=C	≈ 6.4	≈ 203
C≡C	≈ 8.4	–
C–O	≈ 3.3	≈ 335
C–H	≈ 4.2	≈ 288
C=O	≈ 8.5	≈ 149
O=O	≈ 5.08	Corresponding to 245 nm
O–H	≈ 4.8	≈ 258
Si–O	≈ 3.1	–

transfer as well as other chemical reactions of cracked components and the formation of aerosols happen within 10 to 100 ms. The typical retention time of exhaust gas in a radiation field is < 1 s. In any event, the longer the possible time of reaction, the more complete the UV degradation of odorous and in general VOC components in dryer exhaust air (VDI 2441).

The use of UV-units will be within the general scope of exhaust air treatment according to *TA Luft*. However, it must be stated that it is not possible to remove methane from exhaust air with the standard low-pressure UV lamps with wavelengths of 185/254 nm on a sufficient scale. In the case of methane (CH_4), the wavelength of maximum UV radiation absorption is far from the typical wavelength of low-pressure lamps. Therefore, methane is barely affected, despite the low energy value for breaking the first C–H bond. Tridecane ($C_{13}H_{28}$) and tetradecane ($C_{14}H_{30}$), two hydrocarbons occurring at higher concentrations in dryer exhaust air, have wavelengths of maximum absorption at 225 and 160 nm with a lower distance to 185 and 254 nm. These gas components are broken up by UV photons due to the low binding energy of H–C and C–C bonds. Generally, there is still sufficient uptake of radiation energy at wavelengths of $<\pm 20$ nm around the maximum absorption wavelength of individual gas molecules to be used for the unbreaking of bonds.

The practical operation of UV units cleaning dryer exhaust air with a high load of a mixture of VOCs shows deposits of oily substances downstream of the UV reactor at places with heavy air turbulence. These dark brown deposits, presumably produced by polymerization processes, can be removed by frequent CIP water cleaning. In the case of dried layers, ethanol or isopropanol may be used. It is assumed that these deposits are created by side reactions of insufficiently cracked VOC substances and that their intensity depends on the ratio of applied UV radiation to mass flow of VOCs in exhaust air and efficiency of mixing of reacting gas components.

6.14 Condensate of Belt Drying Systems and Its Treatment

6.14.1 Origin and Composition of Condensate

Belt dryers with condensation systems are usually used at wastewater treatment plants where a sufficient capacity of wastewater processing is available. The condensation of water vapor at low temperatures of 35–45 °C improves the thermal efficiency of drying and lowers the specific thermal energy demand for water evaporation. As exhaust airflow is low, the electric energy demand for the exhaust air ventilator is also low. The condensing system facilitates the internal and external use of waste energy. The condensation of water vapor also results in the removal of gas components. A rough estimation of VOC removal in a condensing unit of 40 °C ranges up to 15–25% of VOC entering the condenser. Inorganic gases such as ammonia (NH_3) or hydrogen sulfide (H_2S) with a high to moderate solubility in water are bound in condensate or reduced to elementary sulfur. Individual VOCs in exhaust air are present at comparatively low concentrations in the mg/m^3 region. For a given process temperature in the condenser—usually around 35–45 °C depending on the air temperature level in the belt dryer—the maximum concentration for saturation in gas phase for an individual non-soluble gas component is determined. Only if the actual concentration exceeds the saturation concentration does condensing of this gas component occur, forming a second liquid phase. However, these concentrations are far higher than the occurring concentration for most of the VOCs in exhaust air. For toluene, this concentration is 258.8 g/m^3 and for tridecane 3.8 g/m^3 at a temperature of 40°C. In the case of Toluene this concentration is by far exceeding their encountered concentrations in

Table 6.23 Concentration of gases at saturation point in g/m^3 for a given condenser temperature at 100,000 Pa

Type of gas/vapor temperature of condensing unit	30 °C	35 °C	40 °C	45 °C	50 °C
Water	30.4	39.6	51.2	65.4	83.0
Toluene	171.7	210.7	258.8	317.9	390.6
Tridecane	0.8	1.1	1.6	2.3	3.3
Dimethyl sulfide	1880.3	2219.4	2604.4	3039.3	3527.7

exhaust air of sludge dryers. Thus, the separation of VOC in the condenser is somewhat limited unless the condenser temperature is not set to lower temperatures. This issue can be demonstrated by calculating the concentration c_{VS} of individual non-soluble gas components when saturated in an exhaust airflow at a given temperature T in K, corresponding to the condenser temperature:

$$c_{VS} = \frac{p_{VS} \times M}{R \times T} \tag{6.40}$$

with R as the universal gas constant, $R = 8.314\,\mathrm{J/(mol \times K)}$, M as the molar mass of the gas component in kg/kmol and T as temperature in K. Table 6.23 provides values for water vapor and different VOC for common water vapor condensing temperatures of 30–50 °C.

In exhaust air of belt dryers, only tridecane may reach the indicated concentration of saturation in some cases. Of course, these values are valid for single gas components in exhaust air and do not fully describe the actual conditions in a mixture of gas components. Nevertheless, the influence of the concentration of the gas component on condensation is clearly demonstrated. When water-insoluble gases are separated in a condensing unit only in the gas phase incorporated in the water vapor condensate, they are easily set free as soon as condensate is recirculated by more than 95%, as is common practice, cooled down, and sprayed again into the condenser, as is the case with cross-flow condensing units.

A more remarkable separation of non-condensing VOC can be achieved for direct condensing systems only if almost 100% of the cooling water is flushed off—a process that can be carried out only at wastewater treatment plants with cleaned water from discharge of the plant due to its tremendous cooling water consumption. It can be estimated to be around 22 m^3 per t of water evaporation ($\pm 10\%$) depending on the temperature of the cooling water. The separation of ammonia (NH_3) reaches up to 85% of mass flow entering the condenser without recirculation of cooling water (Engl 2010).

Indirect condensing systems such as finned tube condensers may also separate more non-condensable VOCs, but on its way down, condensate comes into intensive contact with cross-flowing exhaust air and some of the VOCs are stripped off the condensate and transferred back to the exhaust air. A typical value for cooling water consumption is

Fig. 6.26 Cross-flow direct spraying condensing line. (HUBER 2019c)

around $30\,m^3/t$ of condensate (water evaporation) for a temperature difference of 20 K (20/40 °C).

Fig. 6.26 shows a cross-flow direct condensing line arranged beside a belt dryer (HUBER 2019c). The larger horizontal pipeline leaving the condenser conducts exhaust air to the left side, where the scrubbers are pictured. On the right side, the plate heat exchangers are shown, arranged in parallel. Thus, when early blocking of the heat exchanger in operation is observed by a pressure rise, the second unit is switched on, while the first unit is ready for cleaning operations. The cooling water consumption for this cross-flow condenser is around $36\,m^3/t$ of water evaporation with a temperature difference of 20 K (10/30 °C).

Typically, waterflow speed in plate heat exchangers is around $0.6\,m/s$. Without a fouling factor, the heat transfer coefficient reaches around $4000–5000\,W/(m^2 \times K)$ for systems of water/water, and the design is made with a reserve in surface of approximately 8% to compensate for fouling losses in heat transfer (Brenner et al. 1995). Fouling is related to fat, oil, or grease as well as to sulfur deposits and strongly depends on the stabilization of the sludge and on the load of organic substances in wastewater entering the treatment plant.

The following parameters influence the quality of the condensate:

- Drying system and dust concentration in exhaust air with an impact on COD, especially on particulate COD caused by dust transfer to the condensate.
- Sludge temperature during drying with an impact on type and quantity of water soluble VOCs transferred to drying air and consequently to condensate.
- A portion of insoluble VOCs is separated in gas phase with water vapor condensate. However, when recirculated to spraying nozzles in the case of a direct spray condensing system, the gas components might be released again to exhaust air. In this context, condensing systems without partial recirculation of condensate are expected to show a higher degree of separation of VOCs.
- Degree of stabilization of dewatered sludge with an impact on volatile components released to drying air.
- Period between dewatering and drying of sludge with an impact on total and particulate COD
- Fat and protein content of sludge with an impact on COD.
- Ammonium content in dewatered sludge and its pH-value with an impact on ammonium content in condensate.

The COD of the condensate has several origins:

- Fat, oil (lipophilic substances $\approx$ 1500–4700 mg COD/g), and protein not removed by wastewater treatment.
- White oil, paraffinic C_{10}–C_{20} compounds, coming from emulsions ($\approx$ 2700 mg COD/g of active substance) or dispersions ($\approx$ 3600 mg COD/g of active substance) of liquid polymers for dewatering of sludge (Schaum 2016).
- Organic acids such as acetic acid (1067 mg COD/g), propionic acid (1513 mg COD/g), valerian acid (2040 mg COD/g), and butyric acid (1810 mg COD/g).
- Hydrocarbons such as toluene (3085 mg COD/g), styrene (3072 mg COD/g), xylene (3125 mg COD/g), naphthalene (2219 mg COD/g), phenol (2383 mg COD/g) (Meinck et al. 1968).
- Dust of exhaust air collected in the condensing unit forming particulate COD. For stabilized sludge, the COD is estimated to be 1.5–2.5 g/g of DM, for non-stabilized sludge, COD values range from 3.0 to 6.0 g/g of DM (Brautlecht 2000).

Overdosing of liquid polymer may result in a higher COD of dryer condensate. As a rule of thumb, around 0.5 kg of white oil is mixed to 1.0 kg of effective substance for dewatering of sludge. Fig. 6.27 illustrates a condensate from sewage sludge drying.

Condensate of indirect tube condensers has features according to various measurements as presented in Table 6.24 (HUBER 2020/22):

It must be mentioned that for direct spray condensing with one-way cooling water usage, the condensate is correspondingly diluted and the actual concentration of ammonium (NH_4–N) and COD needs to be recalculated.

Fig. 6.27 Condensate from an indirect tube condensing unit. The oil phase on top is clearly illustrated. (Heindl 2018)

A single-belt drying system with a drying air temperature of up to 80 °C, drying digested sludge from a wastewater treatment plant and equipped with a direct spraying condenser, showed condensate concentrations of COD of 1000–2000 mg/L, of BOD_5 of 1000–1500 mg/L, and an ammonium concentration of 1000–2000 mg/L (Dornier 1997).

Table 6.24 Features of condensate from an indirect condenser

Features	Value ranges
COD total	400–1000–4000
COD due to particles—portion of total COD in %	3–10
Ammonium (NH_4^+–N) in mg/L	400–1000
Volatile acids expressed as acetic acid equivalent in mg/L	80–150
Particulate mass filtered by 0.45 µm	1–10
pH	8.5–9.6
BOB_5/COB	0.35–0.70
TOC/COD*	≈ 1/6
Electrical conductivity in µS/cm	2000–3500
Phosphorus in mg P/L	≈ 0.5–2.0
Typical temperature of condensate leaving condenser after air has passed through heat exchanger of internal heat recovery	≈ 40–45 °C
Color	White to yellowish-gray

*Ammonium content accounts for approximately 50% of the chemical oxygen demand (*COD*)

Table 6.25 Different regions of BOD_5/COD ratios after Huber et al. (1993)

Ratio of BOD_5/COD	Biodegradability of organic compounds in wastewater
> 0.59	Easily and entirely biodegradable
0.10–0.59	Incomplete biodegradation
< 0.10	Persistent and non-biodegradable

BOD_5, biological oxygen demand over 5 days; *COD*, chemical oxygen demand

Due to its contents of COD and ammonium (NH_4–N), condensate can be classified as at least "slightly hazardous for water." In Germany, this would correspond to a water hazard class WGK1. Local regulations must be observed, especially in the case of tanks for intermediate storage of condensate.

As a rule, a , biological oxygen demand over 5 days (BOD_5) to COD ratio of more than 0.5–0.6 is related to good biodegradability. Huber distinguished different regions of these ratios according to Table 6.25 (Huber et al. 1993).

However, it is not the only indicator. A Zahn–Wellens test is sometimes performed with regard to biodegradability of condensate. Fig. 6.28 shows such a test (modified from h2ortner 2019). However, when a Zahn–Wellens test for aerobic biodegradability testing is performed with condensate, rapid degradation of COD, i.e., a reduction in COD of more than 20% within 3 h, does not indicate good biodegradability, but is a sign that COD

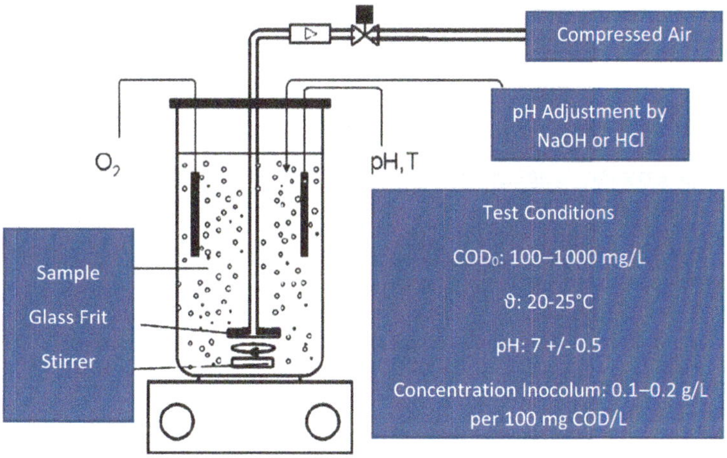

Fig. 6.28 Zahn–Wellens test to analyze biodegradability according to EN ISO 9888 (h2ortner 2022). (Modified from h2ortner 2019)

causing components in condensate have been adsorbed physically to sludge flocs or that other non-biological effects caused a reduction.

Normally, the phosphorus content in condensate is determined by dust generated during the drying process and captured in the condenser. The higher the dust content in air entering the condensing line and the higher the P value in DM, the higher the P content in condensate.

Wastewater treatment plant operators drying their own dewatered sludge on site need to check the recharging of nitrogen when condensate is recirculated. Typically, the recharging of nitrogen due to dewatering of sludge from a municipal wastewater treatment plant with digestion ranges from 15 to 25% or roughly around 1.5 g $N/(P \times d)$, while nitrogen in wastewater flow at the entrance of the plant its assumed to be around 11.0 g $N/(P \times d)$ (Jardin et al. 2006). In Germany, for wastewater treatment plants of more than 100,000 P capacity, the maximum value of treated cleaned wastewater is limited to 13 mg $N_{total,inorganic}/L$.

The nitrogen load calculated by the ammonium content of condensate recirculated to the plant must be added to this value by assuming that around 60–70% of the NH_3 of the dryer exhaust air is loaded to the condensate as NH_4. As already mentioned, the maximum value of ammonium (NH_4–N) in dewatered sludge transferred to dryer exhaust air as ammonia (NH_3) depends on temperature, pH-value, and ammonia content of dewatered sludge and can be estimated according to formula (6.10).

However, in cases where the majority of sludge is taken from external sources, the wastewater treatment plant is at risk of becoming overloaded, at least with regard to nitrogen. Air stripping in a partial flow process as well as biological processes can be used at wastewater treatment sites to remove nitrogen. Nitrification/denitrification processes via a sequencing batch reactor (SBR) with additional carbon source or de-ammonification processes can also be carried out. In addition, at sites where there is no or an insufficient wastewater treatment capacity, condensate processing must be performed due to its high load in ammonium and COD before it can be flushed into the sewer system.

6.14.2 Condensate Treatment

There are two main tasks in the cleaning of condensate before indirect flushing to the sewers is possible:

- Ammonium must be significantly reduced to avoid nitrogen-overloading of the wastewater treatment plant nearby.
- COD must be reduced.

The situation needs to be clarified with the local wastewater treatment plant to obtain the admissible values for ammonium and COD of the pre-cleaned condensate released into the sewage system.

To date (2021), there are no industrial-scale systems for condensate processing with long-term experience and fully complying with all requirements in terms of final condensate quality. In addition, wastewater for the back-flushing of equipment (cleaning process) in the range of 5–20% of the incoming condensate flow need to be repurified or incinerated.

Systems treating condensate must be equipped with a filtration process that removes the substances responsible for particulate COD as a first step. Inclined one-way filters, sand filters, or disc filters can be used.

6.14.2.1 Stripping for Ammonium Removal

If only ammonium needs to be reduced, water vapor or air stripping equipment are used and described in Table 6.26. An air stripping unit is shown in Fig. 6.29 (RVT 2020).

Table 6.26 Systems for ammonium stripping. (Data from offers 2019/2020)

	System	
	Water vapor stripping	Air stripping
Final product	Ammonium water (NH_4OH) (20–22 mass%) to be used for exhaust gas treatment in incineration plants	Ammonium sulfate [$(NH_4)_2SO_4$] (38–40 mass%) to be used as a basic component for fertilizer production
Efficiency of stripping relative to separation of ammonia	Up to 99%	Up to 98%
Production of final product with initial content of NH_4–N of 2000 mg/L	$\approx$ 16.6 kg NH_4OH (21 mass%)/m^3 condensate (99% reduction)	$\approx$ 5.0 kg $(NH_4)_2SO_4$ (39 mass%)/m^3 condensate (98% reduction)
Temperature range	90–95 °C	60–65 °C
Air demand (15 °C)	–	$\approx 1250\,m^3/m^3$ condensate
Steam demand (saturated 4 bar[a])	$\approx 150\,kg/m^3$ condensate (temperature 20 °C) $\approx 80\,kg/m^3$ condensate (temperature 80 °C)	Reheating of exhaust air of column by 20 K: $\approx 14.4\,kg/m^3$
Specific electric energy demand	$\approx 2.0\,kWh/m^3$ condensate	$\approx 1.0\,kWh/m^3$ condensate
Specific condensate throughput	$\approx 5.7\,m^2/(m^2 \times h)$	$\approx 32.5\,m^3/(m^2 \times h)$ relative to cross-section of stripping section $\approx 46.8\,m^3/(m^2 \times h)$ relative to enriching section
NaOH (30 mass%) demand for increasing pH-value from 8.5 to 10 if necessary	$\approx 13\,g/m^3$ condensate	$\approx 13\,g/m^3$ condensate
Cooling water demand for condensing process	$\approx 8.2\,m^3/m^3$ condensate	–

Fig. 6.29 Air stripping unit for ammonium removal. (RVT 2020)

6.14.2.2 Removal of Chemical Oxygen Demand

COD removal can be carried out using, for example, activated carbon filters or a combination of ultrafiltration and reverse osmosis. In any event, one must consider that there will be additional wastewater resulting from back-flushing and cleaning processes or from retentates, ranging from 5 to 20% of the treated condensate stream. This highly loaded residual wastewater is usually combusted.

Ultrafiltration and reverse osmosis

The combination of ultrafiltration and reverse osmosis has the advantage of being able to separate both COD and ammonia (NH_4–N). However, this process is restricted to the limits of the incoming mass flows of FOG (fat, oil, and grease), sulfides and silicates forming deposits. In addition, a positive REDOX potential of condensate increases requirements for back-flushing and cleaning processes. Intermediate cooling of condensate after the ultrafiltration step—causing an elevation in temperature of 3 to 10 K—to temperatures below 40 °C is necessary in some cases and the pH-value must be elevated. The two-stage reverse osmosis system is operated at pressures of 70 and 25 bar absolute. The specific electric energy demand is approximately 11–13 kWh/m^3 of condensate. A pilot test showed a degradation of COD ranging from around 80 to 90%. However, the degradation value to be adjusted depends on the limits of COD and NH_4–N claimed by the client/wastewater treatment plant.

Filtering with activated carbon

When operating several activated carbon filters in series, the filter with the new carbon bed always needs to be flowed through last to exploit the maximum adsorption capacity of activated carbon. However, the maximum adsorption capacity (loads) determined in the 24-h shake test can only be exploited by 70–80% even in sophisticated, staged activated carbon filter arrangements. The maximum adsorption capacity depending on final COD in equilibrium is determined in a 24-h shake test and increases with the initial COD concentration of the condensate. Typical conservative maximum loads—also depending on type and inner surface of the activated carbon used—cover the values of 300–1000 mg COD per g of activated carbon for equilibrium concentrations of COD of 500–3000 mg COD/L at 25 °C. In certain cases, when using activated carbon with an inner surface equal or above 1100 m^2/g and favorable fit of diameter of COD-carrying compounds (typical molecular diameter of 0.4–1.2 nm) with the pore diameter distribution of the activated carbon used (high micro-pore content), maximum adsorption values of more than 1500 mg COD per g of activated carbon might be achieved at initial COD concentrations of condensate exceeding 3000 mg/L.

The specific electric energy demand for activated carbon adsorption of COD ranges from 10 to 11 kWh/m^3 of condensate treated. This figure varies with the limits of COD set in a project. However, one must consider that ammonium is not removed by the activated carbon filter and additional energy consumption for stripping of condensate needs to be added to the energy calculation. Typical activated carbons for wastewater cleaning are described in Table 6.27. It must be noted that wet activated carbon extracts oxygen from ambient air and that filter equipment must be ventilated. Adsorption of water on dry activated carbon also results in a temperature rise: per mole of water, around 286 kJ are released and the water content of carbon is increased up to 40% relative to total mass.

Backwashing of the activated carbon is an important step to remove air bubbles and adherent dust from the activated carbon layer. Thereafter, the filter should be operated upwards with a specific water flow of approximately 15–20 m^3/h per m^2 filter area. To avoid the loss of granules, the flotations of the activated carbon should be low. When backwashing carbon filters, one must consider an expansion of the bed, which decreases with augmenting temperature of wastewater. Also, water filtration resistance decreases with rising temperature, for example, around 200 Pa/m for activated carbon of 8 × 30 mesh for a 5-K temperature rise at a condensate speed of 10 m/h and in the temperature range of 5–25 °C (Jacobi 2022). Before feeding the activated carbon filter, condensate

Table 6.27 Typical activated carbons used for wastewater cleaning. (Donau Carbon 2021; Jacobi 2022)

Features of activated carbon	Hydraffin 30 N (Donau carbon)	Hydraffin AR 8 × 30 (Donau carbon)	AquaSorb 5000 (Jacobi)
Basic material	Bituminous coal	Bituminous coal	Lignite
Granulation	8 × 30 mesh 0.6–2.36 mm	8 × 30 mesh 0.6–2.36 mm	8 × 30 mesh 0.6–2.36 mm
Water filtration resistance in Pa/m	At 13 °C: 5 m/h: ≈ 800 10 m/h: ≈ 1300 20 m/h: ≈ 2600 30 m/h: ≈ 4100	Not indicated	At 15 °C: 5 m/h: ≈ 800 10 m/h: ≈ 1600 20 m/h: ≈ 2500 30 m/h: ≈ 5000
Bulk density in dry condition with < 5 mass% of water content in kg/m^3	490 ± 30	500 ± 30	320
Bulk density after classification by backwashing with water in kg/m^3	≈ 440	≈ 450	≈ 280
Inner BET surface in m^2/g	≈ 950	≈ 1100	≈ 1200
Ash content in mass%	< 15	< 15	< 15
Iodine adsorption in mg/g	1000 ± 50	900	1100
Hardness in mass%	> 90	> 90	Not indicated
Total pore volume in cm^3/g	≈ 0.742	Not indicated	Not indicated
Distribution of pore volume in cm^3/g	Micro-pores 0.2–1.0 nm: 0.366 Meso-pores 1.0–25 nm: 0.143 Macro-pores > 25 nm: 0.232	Not indicated	Not indicated

BET, Brunauer–Emmett–Teller

must be cooled to a temperature of less than 35 °C to limit a decline in adsorption capacity with rising temperature. Typical values of wastewater speed are around 5–10 m/h in free cross-section and a retention time of wastewater in the activated carbon bed of > 30 min.

Fig. 6.30 illustrates the distribution of pores for the activated carbon Hydraffin 30 N and indicates the radius of different molecules occurring in condensate (modified from Donau Carbon 2019). Long-chain FOG occurring in a higher concentration might block meso-pores of activated carbon unless they are not separated by a fine sieving unit in front of the activated carbon filter.

Biological cleaning has also been performed in the past, but frequent fluctuations in condensate constituents—as occurs when drying many different sludge qualities—might overwhelm the adaptive capacity of the microorganism population, as practical experience has shown in some cases.

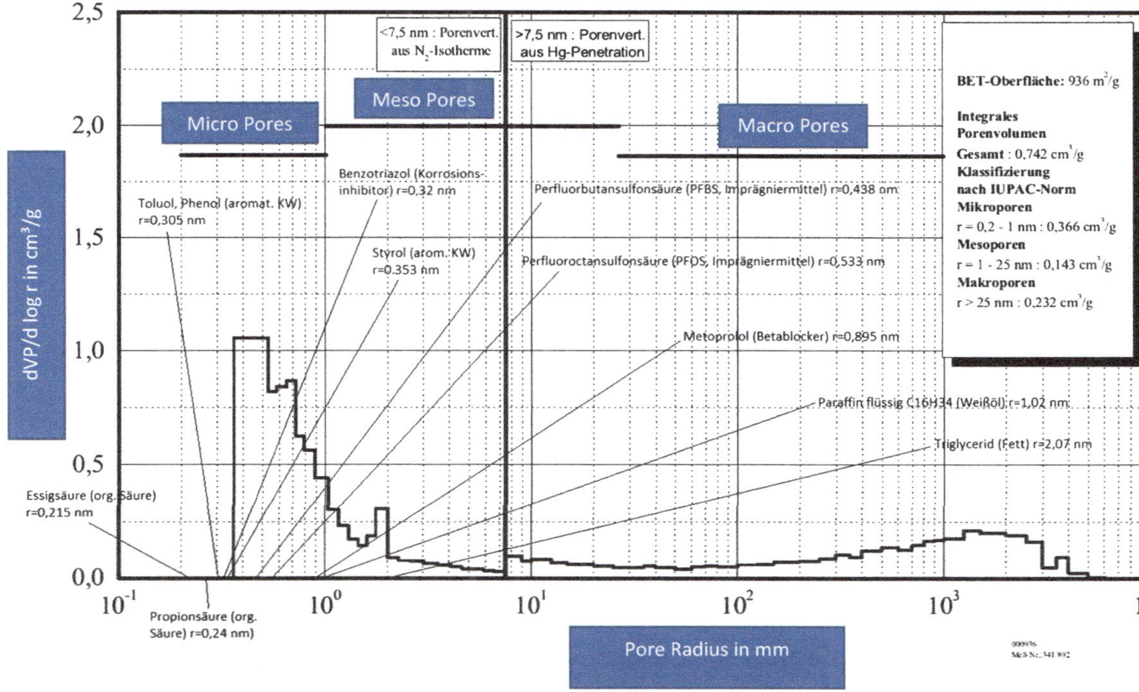

Fig. 6.30 Pore distribution for Hydraffin 30 N activated carbon indicated as pore radius r (*Porenradius*; Donau Carbon 2019) and molecular radius of occurring pollutants in nm (*Mikroporen*, micro-pores; *Mesoporen*, meso-pores; *Makroporen*, macro-pores). *BET-Oberfläche*, *BET* (Brunauer–Emmett–Teller) surface; *Integrales Porenvolumen*, integral pore volume; *Essigsäure*, acetic acid; *Toluol*, toluene; *Phenol*, phenol; *aromat. KW*, aromatic hydrocarbons; *Styrol*, styrole; *Benzotriazol*, benzotriazole; *Paraffin flüssig (Weißöl)*, liquid paraffin (white oil); *Triglycerid*, triglyceride; *Perfluorbutansulfonsäure*, perfluorobutanesulfonic acid; *Perfluoroctansulfonsäure*, perfluorooctanesulfonic acid

Vacuum evaporation

Vacuum evaporation or concentration may be feasible when exhaust energy in a temperature range of 90 °C, for example, warm water, or surplus electric energy is available and there are high demands concerning the final limits of cleaned condensate. The final concentration is limited by the formation of salt crystals.

Two systems are available, and the decision is made, among other influencing factors, on the basis of local energy costs:

- A system with mechanical compression and higher electric energy demand ($\approx$ 50–60 kWh_e/m^3 of condensate)
- A system with thermal energy transfer, lower electric energy demand ($\approx$ 12–15 kWh_e/m^3 of condensate) and thermal energy demand, depending on the number of stages

Of course, electric and thermal energy demand are determined by the degree of concentration, which may be between 3 and 8%, a figure calculated by the final weight of concentrate divided by the initial weight of condensate.

The distilled liquid—generated by condensation—corresponds to the cleaned condensate, while the concentrated liquid remaining in the evaporator needs to be incinerated. This process is able to reduce COD and NH_4–N, but capital expenditure (CAPEX) is high and the application is performed only for smaller evaporation capacities. Depending on vacuum, final concentration, and exhaust heat temperature, the duration of the process is up to 4–6 h at 85 °C. One must consider that:

- The evaporation temperature increases around 7 K towards the end of the process, i.e., from 69 to 76 °C at a vacuum of 300 mbar (absolute).
- The temperature difference between heating agent and boiling condensate consequently decreases and, as a result, the vacuum needs to be reduced from 300 to 250 mbar (absolute).
- Ammonia (NH_4–N) transfer decreases along with pH reduction from 9–10 to around 4–5 in the evaporator and can be controlled by adjusting low pH-values through acid dosing.
- COD degradation of 40–70% can be achieved, depending on initial COD concentration and duration of vacuum evaporation (final concentration).
- NH_4–N degradation of 60–80% is possible, depending on initial NH_4–N concentration, change in pH during treatment, and duration of vacuum evaporation (final concentration).
- Volatile acids are concentrated in the concentrated liquid.
- Thermal energy demand depends on the number of evaporation stages: $\approx$ 750 kWh/t of water evaporation for one stage; $\approx$ 375 kWh/t for two stages; and $\approx$ 250 kWh/t for three stages. The decision on the number of stages is made on the basis of thermal (exhaust) energy costs and investment, which increase considerably with the number of stages.

In all condensate cleaning systems, the cleaned condensate must be cooled to achieve 35 °C as the maximum admissible temperature for indirect flushing into sewers. Depending on local conditions, this can be done by internal heat transfer in the system or by external cooling.

The corrosion issue
It must be emphasized that the condensate from sewage sludge drying may result in corrosion even to 304-grade stainless steel. Practical experience from sites showed that no serious problems with corrosion occurred with 316-grade stainless steel. Therefore, it is recommended that parts of all processing machines or pipes, etc., with direct contact to condensate be made of 316- or 316L-grade stainless steel. EPDM must not be used as a sealing material—fluorelastomers such as Viton must be chosen. Pumps must be equipped with a suitable sealant to prevent diffusion of ammonium (NH_4^+) and odor generation by ammonia (NH_3) in the processing room. Pumps for use in chemical industries with a magnetic drive can also be used.

Total costs of condensate cleaning (2019 price basis)
The total costs are determined by the limits set for COD and ammonium in cleaned condensate and their initial contents. These may range from € 12/m^3 to more than € 25/m^3 of condensate at the end of the second decade of this century. As a rule of thumb, the costs for removal of ammonium by stripping are set at around € 4–€ 5/m^3 of condensate. An economic evaluation must be made in comparison to thermal disposal of condensate, where prices might reach € 20–€ 30/m^3 of condensate and more due to the high energy demand for incineration and transport costs.

6.15 On-Site Measurement of Exhaust Air

Fig. 6.31 shows inline measurement in an exhaust air tube performed by the TÜV (German Technical Inspection Association) Rheinland, Germany. The vertical device is for measuring gas speed and dust content, while the horizontal device is for the chemical analysis of specific gas components such as ammonia, hydrogen sulfide, etc.

Fig. 6.32 shows gas sampling for the evaluation of odor units, performed by HUBER SE in an exhaust air tube between two scrubbers. The plastic bag needs to be filled with ambient air or for more accurate results with dry nitrogen to around half of the cylinder volume and then flushed with exhaust air. Thus, condensing is limited and the odor panel test needs to be corrected with a dilution factor. More exact sampling is achieved by flushing the bag with dry nitrogen and thus thinning the sample to avoid any condensation in the bag and loss of some odor units for evaluation.

The total VOC load and the methane load of exhaust air is measured by a flame ionization detector (FID) or by a photoionization detector PID). When measuring VOC-concentration in exhaust air one has to consider that different gas components have dif-

Fig. 6.31 In-line measurement of exhaust air by the TÜV (German Technical Inspection Association) Rheinland, Germany. (Heindl 2021)

Fig. 6.32 Gas sampling for the evaluation of odor units in exhaust air. (Heindl 2021)

ferent response factors in the FID detection. A surview of response factors is given in the annex. The PID is more suitable for continuous measurements and is based on a UV lamp. However, short chain hydrocarbons < C4 such as methane cannot be detected.

Annex: Selection of Response factors of different gas components in comparison to propane, whose factor is 1.0 (Testa 2023): Acetaldehyd 0.55, acetic Acid 0.52, Aceton 0.76, Methanole 0.74, *n*-Decane 1.00, Tetradecane 1.00, Toluene 1.00, Xylol 1.01.

How gas sampling points have been designed and how gas measurements have been performed is described in EN 15259:2008-1 'Air quality—Measurement of stationary source emissions: Requirements for measurement sections and sites and for the measurement objective, plan and report.' Typically, 3 inch ports are used for air tubes to give sufficient space for all types of gas measurement devices. In the case of negative pressure, units with sufficient suction power need to be used at sealed ports to overcome the vacuum inside.

References

ATV (1994) Exhaust air emissions from municipal wastewater treatment plants (in German). Dokumentation und Schriftenreihe aus Wissenschaft und Praxis. Abwassertechnischen Vereinigung e. V., Hennef

Bank M (1995) Basic Knowledge Environmental Technology (in German), 3rd edn. Vogel, Würzburg

Bashova S, Bagreev A, Bandosz TJ (2003) Adsorption/oxidation of methyl mercaptan on activated carbons. Prepr Pap-am Chem Soc Div Fuel Chem 48(2):690

Bathen D, Breitbach M (2001) Adsorption Technique (in German). Springer VDI Verlag, Berlin Heidelberg

Bratzler K (1944) Adsorption of Gases and Vapors (In German). Theodor Steinkopff, Dresden Leipzig

Brautlecht P (2000) Technical and Economic Aspects of Drying Plants for Municipal Sludge (In German). Diss. D82 TH AachenBrenner A et al. (1995) Plate Heat Exchangers (In German). Verlag Moderne Industrie. Die Bibliothek der Technik 105. Landsberg/Lech

Brenner K et al (1995) Plate Heat Exchanger (In German). Verlag Moderne Industrie. Landsberg/Lech

Bürkholz A (1989) Droplet Separation. VCH Verlagsgesellschaft, Weinheim

Claus H (2021) Ozone Generation by Ultraviolet Lamps. Photochem Photobiol 97:471–476

Data from offers (2019/2020) Technical data on air and steam stripping systems

De Bruyn WJ et al (1995) Henry's Law Solubilities and Setchenow Coefficients for Biogenic Reduced Sulfur Species Obtained from Gas Liquid Measurements. J Geophys Res 100(D4):7245–7251

Devos M et al (1990) Standardized Human Olfactory Thresholds. IRL Press at Oxford University Press (Reprinted 2011)

Dobslaw D (2021) Photo of GC-MS device, University of Stuttgart

Dobslaw D (2022) Personal Communication. Head of Institute for biological exhaust air treatment. ISWA University of Stuttgart

Dobslaw D (2023) Personal communication. TTI GmbH ALR Umwelt TGU Weil der Stadt. Germany

Dobslaw D Wilde C Engesser KH (2010) Possibilities of Exhaust Air Treatment of Emissions from Sewage Sludge Drying (in German). VDI-Berichte 2110, 237–42 Düsseldorf

Donau Carbon (2003) Technical Data Sheet Desorex G50, e1812/03. Frankfurt a.M.

Donau Carbon (2019) Diagramm on Pore Size Distribution of Hydraffin 30 N activated carbon. Frankfurt a.M.

Donau Carbon (2020) Data on Adsorption of Toluene on Activated Carbon. Frankfurt a.M.

Donau Carbon (2021) Technical Data Sheets Hydraffin 30 N and AR 30x8. Frankfurt a.M.

Donau Carbon (2022) Pore Size Distribution of Activated Carbon UC 50 on coconut basis. Frankfurt a.M.

Dornier (1997) Decentralized drying of sewage sludge (In German). CAV 1:12–14

Dürr H (1975) Photochemistry at the Carbon-Sulfur-System (In German). In: Houben-Weyl Methods of Organic Chemistry IV/5b Photochemistry II. Thieme, Stuttgart

Dürr (2022) Personal Communication on RTO (www.durr.com)

Ecochimica (2010) Prospectus Bio Filters and Bio Modules. Technical Report on the Ecochimica Bio Filter. VER 1.2-MAG/10. Creazzo

Elektrische Glühlampen mbH (2008) Mercury Amalgam for a Discharge Lamp and a Discharge Lamp. Patent WO 2008/017654 A1 14th of February 2008

Ellis C, Wells AA (1941) The Chemical Action of Ultraviolet Rays. Reinhold Publ

Engl K (2010) Storia dell'essicatore e termovalizzatore presso ARA Tobl (in Italian language). ARA Pustertal

Fehnel EA, Carmack M (1949) The ultraviolet absorption spectra of organic sulfur compounds. I. Compounds containing the sulfide function. J Am Chem Soc 71(1):84–93

Franke W (2011) Bio Filter as a Component of Combined Exhaust Air Treatment Processes in Wastewater Industry (in German). Wasser Abwasser Umwelt Schriftenreihe des Fachgebietes Siedlungswasserwirtschaft, vol 32. Universität Kassel

h2ortner (2019) https://www.h2ortner.com/fileadmin/Media/Produktinformationen/Information_Zahn-Wellens-Test.pdf

h2ortner (2022) Publication of h2ortner on Zahn-Wellens-Test

Heindl A (2010) Photo of Filling a Biofilter with Wood Chips

Heindl A (2012) Bio filter of wastewater treatment plant in Balingen, Germany

Heindl A (2015) Practice book of belt drying – Basics, Applications, Calculations (In German). Springer, Heidelberg

Heindl A (2018) Photo of Condensate from an Indirect Tube Condenser Applied in Belt Drying of Sludge

Heindl A (2020) Photo of a pair of chemical scrubbers for exhaust air treatment of a belt dryer with condensing system

Heindl A (2021) Photo of Exhaust Air Measurements

Heindl A (2022) Photo of a Covered Biofilter

Heraeus Noblelight (2021) Ultraviolet Light for Water Treatment. Technical Handbook. Hanau

Heraeus Noblelight (2023) Personal Communication on amalgam lamps and cleaning agents for quartz glass. Hanau

Huber S, Huber W, Frimmel H (1993) Summary Parameters for Organic Compounds. In: DVGW Water Chemistry for Engineers (In German). R. Oldenbourg, Munich

HUBER (2019a) Photo of Deposits on a Nozzle of an Acid Scrubber. Berching

HUBER (2019b) Photo of a condensate. Berching

HUBER (2019c) Photo of condensing line of HUBER belt drier in Innsbruck

HUBER (2020/22) Analysis of dewatered sludge with GC-MS headspace and SPME

Jacobi (2022) Technical Data Sheet TDS-Aquasorb-5000-A4-GER-B1016. Kalmar

Jardin N et al (2006) Recharging From Sludge Treatment – Processes of Treatment of Sludge Water (In German). Report of the DWA Working Group AK 1.3 Recharging from sludge Treatment. https://www.araconsult.at/download/literature/atv_jardin_wurrzburg.pdf. Accessed 21-08-2022

Knop W et al (1972) Techniques of Keeping Air Clean (In German), 2nd edn. Otto Krausskopf, Mainz

Kopp J (2006) Examinations on sludge stability and odor development in case of sludge drying at wwtp Ingolstadt (in German). Internal report for HUBER SE. Lengede

Lange F (1971) Technical and Economic Considerations on Removal of Ammonium Gas from Coke Oven Gas (In German). Doctor work RWTH Aachen

Likusta (2022) https://www.likusta.com/de/produkte/abluftbehandlungsanlagen/biowaescher-und-biotricklingfilter/. Accessed 23 Aug 2022

Margesin R, Schneider M, Schinner F (1996) Practice of biotechnological exhaust air cleaning (In German). Springer, Berlin Heidelberg

Meinck F, Stoof H, Kohlschütter H (1968) Wastewater of Industry (in German), 4th edn. Gustav Fischer, Stuttgart

Michaud M (2022) Calculation of Solubility of Methane Thiole in Dependance on pH-Value and Temperature (personal communication)

Nagata Y (n.d.) https://www.env.go.jp/content/900450144.pdf

PCC industries (2022) https://ppcair.com/products-services/biofiltration/biofiltration-factors

Rant Z (1977) Evaporation in Theory and Practice (in German), 2nd edn. Sauerländer, Aarau Frankfurt a.M

Reichert K et al (1998) *Pseudonocardia asaccharolytica sp. nov.* and *Pseudonocardia sulfidoxydans sp. nov.*, two new dimethyl disulfide-degrading actinomycetes and emended description of the genus Pseudonocardia. Int J Syst Bacteriol 48(Pt 2):441–449

Reid RC, Prausnitz JM, Sherwood TK (1977) The Properties of Gases and Liquids. McGraw-Hill, New York

Reiser M, Fischer K (1996) How do high performance bio filter stand the test under difficult conditions in practical operation? In: Margesin R et al (ed) Practice of biotechnological exhaust air cleaning. Springer, Berlin Heidelberg

Rutenfranz C (1996) Influence of Air-Water-Distribution Balance of Harmful Substances and further Parameters on the Kinetics of Biological Exhaust Air Treatment Processes (In German). VDI Fortschrittsberichte Reihe 15: Umwelttechnik. VDI, Düsseldorf

Ruth JH (1986) Odor thresholds and Irritation Levels of Several Chemical Substances: A Review. Ind Hyg Asso J 47(3):A142–A151

RVT (2020) Photo of an air stripping system for ammonium removal. Marktrodach

Schaum C A (2016) Future Wastewater Treatment. Protection of Health, Watersides and Resources (In German). Habilitation TU Darmstadt D17

Schmidt J (1970) Processes of Gas Preparation (in German). VEB Deutscher Verlag für Grundstoffindustrie, Leipzig

Schön M, Hübner R (1996) Odor – measuring and elimination (in German). Vogel, Würzburg

Schweisfurth R (1988) Applied Microbiology of Hydrocarbons in Idustry and Environment (in German). Expert Verlag Ehningen, Germany

Semjonow NN (1961) Some Problems of Chemical kinetics and ability of reaction (In German). Akademie, Berlin

Shareefdeen Z, Singh A (2005) Biotechnology for odor and Air Pollution Control. Springer, Berlin Heidelberg New York

Siedlewski J (1961) The properties and Structure of Activated Carbon in Catalytic Processes (In Polish Language). Studia Societatis Scientiarum Torunensis, Torun

Störk (2022) Prospetcus Belfor Bio Filter Material. Werk 38, Emmingen

Stuetz R, Frechen F-B (2001) Odours in Wastewater Treatement – Measurement, Modelling and Control. IWA Publishing London

Summer W (1962) Ultra-Violet and Infra-Red Engineering. Sir Isaac Pitman & Sons, London

Summer W (1970) Making air and wastewater odorless (In German). R. Oldenbourg, München

Täffner T (2011) Adsoptive separation of harmful substances from complex exhaust gases for their application in exhaust air treatment (in German). Diss. Faukultät für Maschinenbau der Universität der Bundeswehr Hamburg

Technische Anleitung zur Reinhaltung der Luft (2021) TA Luft German regulation for keeping air clean

Tejada Gutierrez G (2011) Examination of drying behavior of sewage sludge (In German). Practical Report during work at HUBER SE. Berching

Testa (2023) Response factors for FID application. Testa GmbH, München, Germany. https://www.testa-fid.de

Tholander (2022) Picture of a crossflow scrubber

Timofejew TP (1967) Kinetics of Adsorption (In German). VEB Deutscher Verlag für Grundstoffindustrie, Leipzig

Trenkwalder J (2011) Using Exhaust Energy for Sludge Drying and Co-Incineration in a Cement Works (In German). 7th DWA Sludge days, Fulda, 29.03.–31.03.2011

Van der Meer M, Van Lierop F, Sokolov D (2023) The analysis of modern low pressure amalgam lamp characteristics. https://uvlampconsulting.com/wp-content/uploads/2017/08/the-analysis-of-modern-low-pressure-amalgam-characteristics-fred-van-lierop.pdf. Accessed 3 Feb 2023

VDI 2441 Process Gas and Waste Gas Cleaning by Cold Plasma (In English/German). Series VDI-Manual Cleaning of Air Vol 6 Exhaust Gas Cleaning Technique. Düsseldorf May 2016

VDI 3477 (2004) Biological waste gas treatment (In English / German). Series VDI-Manual Cleaning of Air Vol 6 Exhaust Gas Cleaning Technique Düsseldorf November 2004. Withdrawn

VDI 3477 (2016) Biological waste gas treatment (In English/German). Series VDI-Manual Cleaning of Air Vol 6 Exhaust Gas Cleaning Technique. Düsseldorf March 2016

VDI 3478 (1996) Biological Waste Gas Purification (In English/German). Series VDI-Manual Cleaning of Air Vol 6 Exhaust Gas Cleaning Technique. Düsseldorf July 1996

VDI 3674 (2013) Waste Gas Cleaning By Adsorption (In English/ German). Düsseldorf April 2013

VDI 3679 (1999/2005) Wet Separators. Waste Gas Cleaning by Adsorption (Scrubbers). (In English and German) Series VDI-Manual Cleaning of Air Vol 6 Exhaust Gas Cleaning Technique. Düsseldorf March 1999/December 2005 (approved again)

Von Kienle H, Bäder E (1980) Activated Carbon and its Industrial Application (In German). Ferdinand Enke, Stuttgart

Voronov A (2008) New Generation of Low Pressure Mercury Lamps for Producing Ozone. Ozone Sci Eng 30:395–397

Wagner H (2012) Influence of temperature on electric conductivity of thinned aqueous solutions (In German). VGB Power Tech 3:3–10

Wang C et al (2009) Effects of Adding Ozone on a Biofilter Treating Gaseous Chlorbenzene. Performance Change and Mechanism. Biotechniques for Air Pollution Control. Proceedings of the 3rd International Congress of Biotechniques For Air Pollutin Control, Delft. CRC Press

WEF (2020) Water Environment Federation WEF Manual of Practice: Odor emissions and Control. Alexandria, USA

Wetter C, Thiel A (2018) Development of a processing unit for hygienisation of pasty biosolids (in German). AZ 33083-31/11. Deutsche Bundesstiftung Umwelt

Worch E (2015) Hydrochemistry – Basic concepts and exercises. De Gruyter, Berlin, Boston

Yaws CL (2015) The Yaws Handbook of Vapor Pressure. Antoine Coefficients, 2nd edn. Elsevier, Amsterdam

ZKA Ingolstadt (2020) Brochure on Drying Plant of Zentrale Kläranlage Ingolstadt, Germany

Zin RM (2017) A New thermodynamic Correlation for Apparent Henry's Law Constants, Infinite Dilution Activity Coefficient and Solubility of Mercaptans in Pure Water. J Nat Gas Eng 2(2):148–170

7 Sludge Analysis with Regard to Belt Drying and Exhaust Air Treatment

For the correct design of a belt drying system, tests on dewatered sludge samples are essential. Extrusion and drying behavior need to be evaluated and the gas components in the exhaust air need to be determined.

How sludge analysis can be performed with regard to gas components transferred to exhaust air and drying behavior is demonstrated for the example of a digested sludge at a German wastewater treatment plant.

Table 7.1 shows the results of chemical and gas chromatography–mass spectrometry (GC-MS) analysis of a dewatered sludge and comments on the effects on belt drying and exhaust air treatment. A digested, dewatered sludge sample was taken in spring from a wastewater treatment plant with a capacity of 360,000 population equivalent (PE). A deammonification process and high-load digestion are carried out with a total digester capacity of 8900 m^3 ($2 \times 700\,m^3$ [high load] $+ 3 \times 2500\,m^3/h$) with a production of sludge of around 4200 metric tons of dry mass (DM) per year, resulting in biogas production of

Table 7.1 Sludge analysis and comments on belt drying and exhaust air treatment

Sludge property of wet chemical analysis of dewatered sludge	Value	Comments on belt drying and exhaust air treatment
Dry residue (DR) in %	26.0	Typical for digested sludge
pH	8.4	High, but still typical for sludge digested in high digesters after stripping of CO_2 during dewatering; most of the ammonia (NH_3) will be transferred to drying air
Total Kjeldahl nitrogen (TKN) in mg/kg DM	56,531	
Protein value = TKN × 6.25/DR in %	29.1	Low, in combination with low VDR → No sticking during belt drying to be expected
Volatile dry residue (VDR) in %	55.7	Low, indication of good digestion

A. Heindl, *Belt Drying of Sludge*, https://doi.org/10.1007/978-3-662-66512-1_7

Table 7.1 (Continued)

Sludge property of wet chemical analysis of dewatered sludge	Value	Comments on belt drying and exhaust air treatment
Ammonium (NH_4–N) in mg/kg DM and in %	15,192 1.52	High value, together with high pH-value, ammonia (NH_3) in dryer exhaust air will be very high
Sulfate sulfur (SO_4–S) in mg S/kg DM and in %	738 0.07	Low value
Total sulfur in mg/kg DM and in %	11,000 1.10	Medium content
Organic sulfur as difference of total sulfur and sulfate sulfur in mg/kg DM and in %	10,262 1.03	High, showing a high potential for organic sulfur components in exhaust air
Volatile organic acids, expressed as acetic acid equivalent in mg/kg DM	1140	Low, indication of good digestion

Fig. 7.1 Dewatered extruded strains of tested sludge. (HUBER 2021)

around 3,000,000 m^3 per year. The degradation of volatile dry mass (VDM) ranges from 54 to 58%—evidence of a very good stabilization process.

Fig. 7.1 shows extruded 10-mm strains of this digested sludge.

The sludge strains were endless and the bed of strains did not flow under gravity. Therefore, porosity of the bed will be sufficient for an even air flow and drying will be fast. The drying curve needs to be monitored in an experimental drying system operated at the same parameters such as load of dewatered sludge in kg per m^2 of drying area, air temperature, air speed through the bed, and absolute humidity of drying air. Of course, there will be a conversion factor to scale up drying time from the test to the industrial drying process. Fig. 7.2 illustrates the drying curve of the examined digested sludge.

The sludge strains shrink during the drying process, as shown in Fig. 7.3.

Headspace analysis of the sludge sample was carried out at 125 °C (283 °F), and Table 7.2 shows the components to be expected in exhaust air from the belt dryer. Headspace analysis focuses on organic sulfur components contributing to odor units in exhaust air.

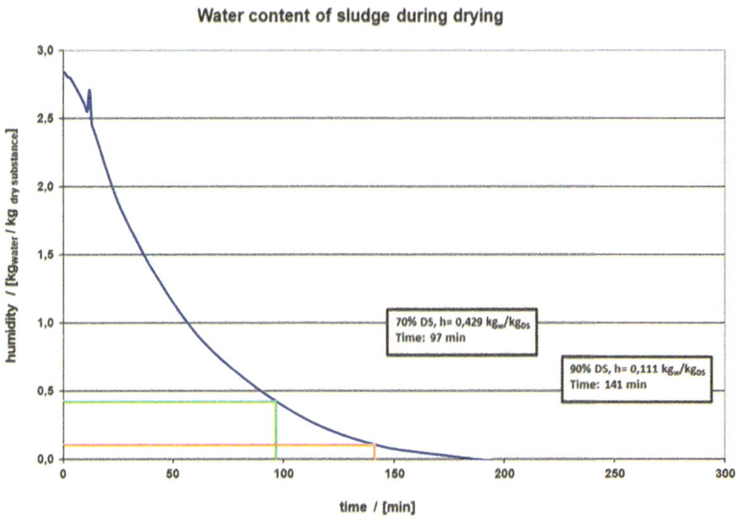

Fig. 7.2 Drying curve of examined dried sludge for an air temperature of 125 °C. (HUBER 2021)

Fig. 7.3 Dried extruded strains of tested sludge. (HUBER 2021)

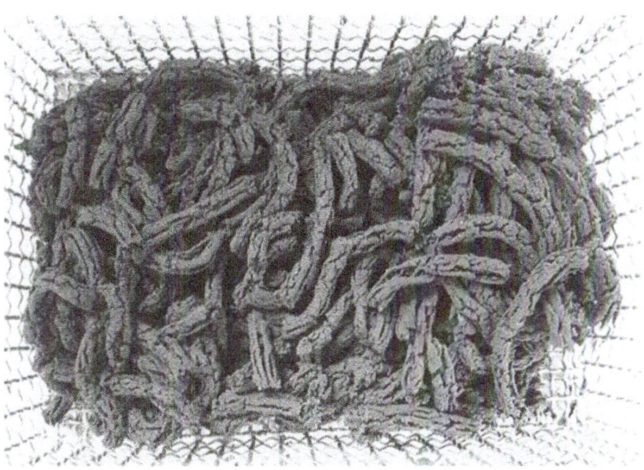

Table 7.2 Gas chromatography–mass spectrometry headspace analysis of a digested dewatered sludge sample with gas components having a mass% portion of more than 1%

Number	Gas component	Mass%
1	Dimethyl disulfide (DMDS) [$(CH_3)_2S_2$]	8.677
2	Tetradecane	6.911
3	Dimethyl disulfide (DMDS) [$(CH_3)_2S_2$]	6.813
4	Dimethyl sulfide (DMS) [$(CH_3)_2S$]	6.607
5	Dimethyl sulfide (DMS) [$(CH_3)_2S$]	5.786
6	Methane thiol (CH_4S)	5.138
7	o-Cymene	4.589

Table 7.2 (Continued)

Number	Gas component	Mass%
8	Methane thiol (CH_4S)	4.489
9	4-(2-Chloroethoxy)-N,N-diethyl-6-methylthio-1,3,5 triazin-2-amine	3.895
10	Pyridin-4-aldehyde	3.367
11	4-Chloro-1,5-naphthyridine-1-N-oxide	3.298
12	4-(Pyrolidin-1-yl)-2,1,3-benzoxadiazole	2.196
13	Cyclohexane, 1-methyl-4-(1-methylethylidene)	2.040
14	Dodecane, 2,6,10-trimethyl-	1.845
15	Pentadecane	1.600
16	Butanale, 3-methyl-	1.552
17	Heptyl-cyclohexane	1.550
18	2,6,10-Trimethyl-tridecane	1.530
19	1,2,5-oxadiazol, 3-ntro-4-(4-nitro-1,2,5-oxadiazol-3-yl)	1.252
20	1,1′-Bicyclohexyl	1.224
21	cis-2-Ethylbicyclo(4.4.0)decane	1.107
22	Octatriacontyl trifluoroacetate	1.098
Sum	Methane thiol (MT, CH_4S)	9.627 (high)
Sum	Dimethyl sulfide (DMS) [$(CH_3)_2S$]	12.393 (high)
Sum	Dimethyl disulfide (DMDS) [$(CH_3)_2S_2$]	15.490 (very high)
Ratio	DMS/MT	1.287 elevated, medium period of intermediate sludge storage after de-watering
Ratio	DMDS/MT	1.609 elevated, some mechanical handling of dewatered sludge with oxygen access to bulk

Fig. 7.4 illustrates the GC-MS evaluation of the digested sludge at a temperature of 125 °C.

The final evaluation of this digested sludge can be described as follows:

- High contents of organic sulfur are to be expected in exhaust air and special attention must be paid to the adjustment of the caustic scrubber.
- Prolonged storage of this sludge will result in the formation of higher exhaust air load of dimethyl sulfide (DMS) with low water solubility.
- Mechanical handling with frequent oxygen access to the bulk of dewatered sludge will generate a higher dimethyl disulfide load in exhaust air with low water solubility.
- Low volatile dry residue (VDR) and low content of volatile acids in the VDM of sludge indicate good digestion and a high degree of stabilization; the caustic scrubber will not be loaded with high amounts of volatile acids.

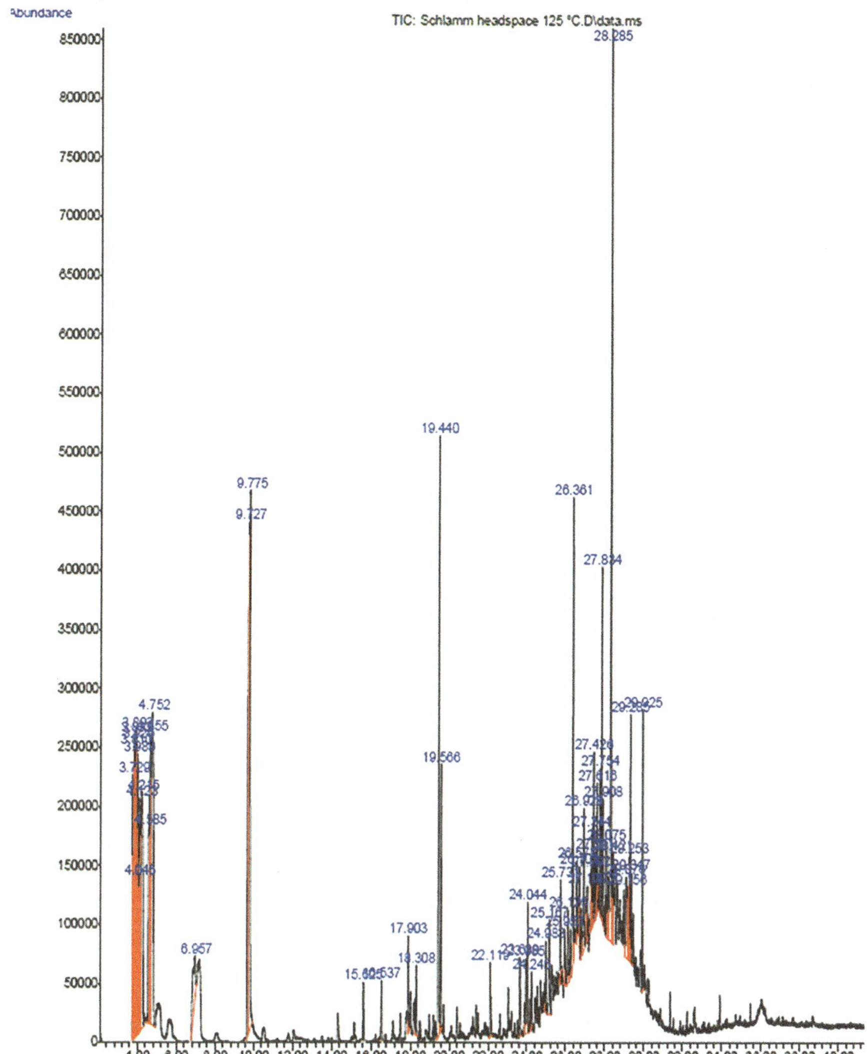

Fig. 7.4 Gas chromatography–mass spectrometry analysis of the digested sludge at 125 °C. (HUBER 2021)

- High ammonia content together with high pH values of dewatered sludge will create a high NH_3 concentration in exhaust air and a high ammonia load in condensate in the case of a condensing system and high consumption of sulfuric acid for exhaust air treatment.
- A low protein value in combination with low VDR does not cause problems with sticking during belt drying.
- The drying process in a belt dryer will be even and fast.

References

HUBER (2021) Full Analysis of a Digested Sludge Together with Drying Curve and Fotos of Sludge in Different Stages of Processing. Berching

8 Storage of Dewatered Sludge

The storage volume for dewatered sludge is calculated with a minimum of 3 days of continuous operation of the belt dryer to be able to bridge prolonged holidays. This decision determines the methane and odor emission of the bulk of dewatered sludge with the corresponding consequences on the design of the exhaust air treatment system according to local regulations. Emissions will be far higher at elevated sludge temperatures during the summer.

There are two main types of sludge storage facilities. A distinction is made between bunkers and silos.

Receiving bunkers are usually made of concrete and located underground for easy feeding of the sludge by trucks. Concrete bunkers have a storage volume of around 500–5000 m^3 depending on drying capacity of the drying plant. Steel bunkers usually have a low volume of 30–100 m^3 and are installed above ground. Fig. 8.1 shows a bunker of this kind (Huning 2022).

The larger bunkers are filled by trucks and emptied by hydraulic walking floors and screw conveyors. The sliding bars have a triangle shape, the vertical side of which is used for sludge conveying. The inclined side of the bars crosses the sludge bulk below. Smaller bunkers are often filled by cranes from a large receiving bunker. The surface to volume ratio of bunkers and their footprint is higher in comparison to cylindrical silos and typically ranges from 0.3 to 1.0 m^2 of sludge surface exposed to headspace per m^3 of sludge. The total discharging of a bunker is not possible—a small amount of dewatered sludge will stay at the bottom of the bunker. First-in-first-out operation is practiced only in the case of filling from the top and discharging via a sliding floor and discharging screw conveyor. Discharging by crane always results in first-in-last-out operation. The latter variant results in reduced mechanical stress on dewatered sludge, thus avoiding cell rupture and related deterioration of sludge quality.

A. Heindl, *Belt Drying of Sludge*, https://doi.org/10.1007/978-3-662-66512-1_8

Fig. 8.1 Steel bunker with hydraulic walking floor and discharging screw. (Huning 2022)

Silos for receiving dewatered sludge are vertical cylinders made of steel and filled from the top and discharged from the bottom by screw conveyor batteries covering the whole cross-section or via a gliding frame with a central screw conveyor. Their diameter ranges from 2 to 12 m, and their height is up to 20 m. The maximum volume is set to 2500 m^3. Their sludge surface to sludge volume ratio ranges between 0.05 and 0.20 m^2/m^3. Fig. 8.2 shows a gliding frame for discharging dewatered sewage sludge at the bottom of a cylindrical silo (SAXLUND 2022) with a flat floor. Typical electric energy consumption for a hydraulic sliding floor system in combination with a discharging twin screw conveyor is around 0.5 kWh per m^3 of discharged dewatered sludge with a DR of 25–35%.

Fig. 8.2 Sliding frame for discharging dewatered sewage sludge stored in a vertical cylindrical silo. (SAXLUND 2022a)

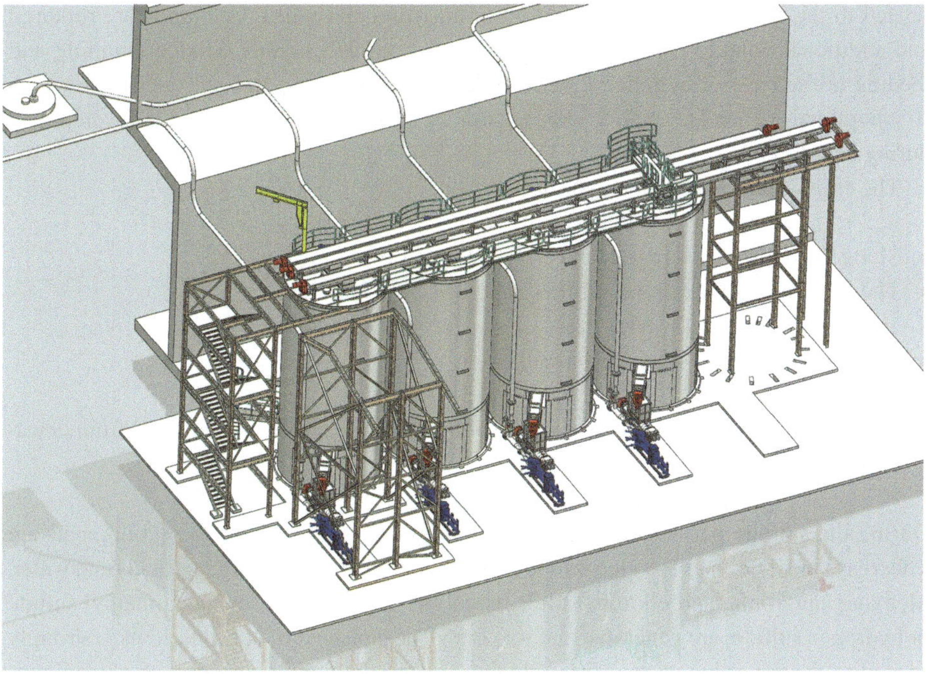

Fig. 8.3 Silo group. (Saxlund 2022b)

Silos have the advantage of a smaller footprint and lower gas emissions of methane due to the smaller surface to volume ratio of sludge. Fig. 8.3 illustrates a silo group and Fig. 8.4 piston pumps for silo filling. However, as a rule, investment is higher for the same sludge volume due to the more expensive sludge feeding facilities. In addition, pumping with piston pumps, which is usually carried out in the case of silos, diminishes sludge quality with regard to greater volatile organic compound (VOC) emissions into the exhaust air of dryers. This effect has been observed even at lower pressure pumping and might be

Fig. 8.4 Piston pumps for silo filling. (Saxlund 2022b)

related to the alternating pressing and tensioning on cell tissues. Cell walls are ruptured and additional volatile carbon acids are formed from cell liquids. Sludge pumping and feeding at the base of cylindrical vertical silos creates first-in-last-out operation, which is suboptimal for sludge quality with regards to sludge structure and odor development during belt drying.

The emission of gas components depends on:

- Sludge mass in the storage facility
- The surface to volume ratio of the sludge bed
- Volatile dry residue (VDR)
- The type and degree of sludge stabilization
- Sludge temperature
- The amount of mechanical stress exerted on the sludge and the time period from dewatering to drying

During the storage period, microbial degradation of organic matter is ongoing, causing a decrease in VDR and pH-value through the formation of volatile organic acids and water. Methane, ammonia, and odorous sulfur-based gas components, such as dimethyl sulfide or hydrogen sulfide, are generated and released. The emission of odorous gases is strongly related to the pH-value of dewatered sludge. During the winter period, with sludge temperatures in between 5 and 10 °C, the emission of gases is limited, but during the summer, with rising sludge temperatures of 15–25 °C, far higher gas emissions—especially in the case of methane—are encountered. Also, the methane content in dryer exhaust air is increased.

The odor emissions from dewatered sludge were measured for 14 German wastewater treatment plants, whereby 34 odor samples were taken (Frechen 2004). The average odor value expressed as geometric mean (arithmetic mean of log) was calculated to be around 2500 OU/(m^2 of surface of dewatered sludge × h), with maximum values exceeding 100,000 OU/(m^2 × h).

Dried sludge plates occurring in the bulk of dewatered sludge are a clear sign that the sludge mass has been stored for a long period of time. These agglomerates must be removed by a special separator to avoid mechanical problems during transport from the storage facility to the dryer. In addition, as already described, a higher mass transfer of VOCs into drying air and consequently dryer exhaust air is to be expected.

Dewatered sludge of good quality results in lower operating costs with regard to maintenance and expenditure for exhaust air treatment. Therefore, the disposal costs for a wastewater treatment plant delivering sludge of this kind should be considerably lower.

By measuring the gas emissions of a 30-m^3 steel bunker for dewatered digested sludge, the effect of sludge temperature and fissures in the sludge bed caused by material discharge at the bottom of the bunker on the amount and type of gas components could be evaluated. The dry residue (DR) of the dewatered sludge was 26.9% and the VDR reached 55.6%. The gas concentration was analyzed with a gas aspirator bulb at the surface of the sludge bed; the measuring device was a DRÄGER X-am 7000. The specific exhaust air flow was determined to be 62.5 $m^3/(m^2$ of bed surface $\times$ h). Under static bed conditions, methane, carbon dioxide, and ammonia could be measured. As soon as sludge was discharged by a cross-conveying screw, carbon monoxide and hydrogen sulfide from the inner volume of the bed could be detected. The measurements were taken during the summer at extreme sludge temperatures of 29 °C and ambient temperatures of more than 30 °C. A very high average value for methane emissions of 0.87 NL CH_4/(kg VDR $\times$ h) was calculated, showing the influence of sludge temperature.

The average annual sludge temperature is around 15 °C, and in the case of larger drying projects, bunkers are located underground and are not exposed to direct sun radiation. This limits the heating up of the sludge.

Fig. 8.5 summarizes specific methane emission measurements in industrial units, calculated in NL/(kg VDM $\times$ h) and related to sludge surface (exposed to headspace) to sludge volume ratio. As a rule of thumb, as the ratio of surface to volume increases, specific methane emissions increase, ranging from ~0.2 to 0.5 NL CH_4/(kg VDM $\times$ h). Extreme values are also marked, which were caused by fissuring of the sludge surface. Gas bulbs of methane discharge from the interior of the sludge bed. This might be the reason for the extreme value of the sludge silo at the Weiher II power plant, Germany (0.08 m^2/m^3 and 1.4 NL CH_4/[kg VDM $\times$ h]). Such incidents are very seldom and of short duration. In this case, the exhaust airflow needs to be automatically increased. In general, slim vertical and cylindrical silos with a low surface to volume ratio generate less specific average methane emissions in comparison to flat bunker units (Heindl and Dobslaw 2022).

During storage of dewatered sludge, different residual degradations of organic dry mass are possible depending on the degree of stabilization:

- Sufficiently stabilized digested sludge: additional degradation of approximately 15% of fed and easily degradable organic dry mass
- Partially stabilized sludge: additional degradation of approximately 30% of fed and easily degradable organic dry mass.

Therefore, dewatered sludge deposited for more than three months will generate far less or practically no methane due to the fact that a diffuse release has already occurred. Detailed information on sludge stabilization can be found in the publication by the working group of the German Association for Water, Wastewater and Waste (DWA), entitled KEK 2.1 Stabilization and Decontamination (DWA 2020).

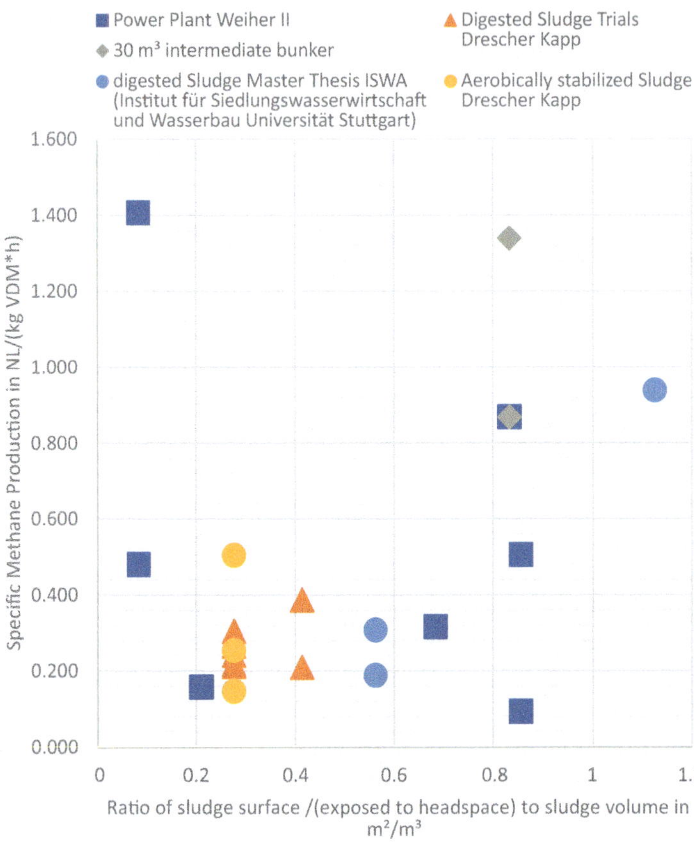

Fig. 8.5 Measured value of methane production in bunkers and silos used for storing dewatered sludge depending on the ratio of sludge surface (headspace) to sludge volume. (Heindl and Dobslaw 2022). *FS* digested sludge; *AS* aerobically stabilized sludge

Each drying plant has two emission points: one for the belt drying system and one for the storage facility of dewatered sludge. Mixing bunker exhaust air with dryer exhaust air results in an increased concentration of methane in the mixed air. Methane can only be removed by thermal oxidation, thus forcing the operator to use a regenerative thermal oxidation (RTO) unit if both exhaust air streams need to be treated in one combined emission point. In this case, the bunker exhaust air must be heated up and introduced into the main exhaust air stream immediately before the RTO unit. Due to the high methane content of bunker exhaust air, the natural gas demand for additional heating of the the RTO unit is reduced. In the future, excimer VUV (vacuum ultraviolet) lamps at the 172 nm wavelength might be used for cold oxidation of methane, provided an economical system of operation can be set up.

High methane emissions from storage facilities for dewatered sludge may have significant impacts on the environment. In addition to the technical solutions mentioned, a process-integrated solution of the methane issue can be found in intelligent logistics management for dewatered sludge. Sufficient sludge stabilization in combination with immediate drying of sludge after dewatering, thereby avoiding long intermediate storage periods, could help reduce diffuse methane emissions. Here, a cooperation between wastewater treatment plants and drying plants is beneficial.

According to BGR 104, the minimum temperature-dependent exhaust air flow, based on a temperature of 20 °C, can be calculated under the condition of constant methane emission from the bed of dewatered sludge:

$$V_{\text{exhaust air}} = G \times f \times \frac{T}{293} \times \frac{1}{(k_{\text{adm}} \times \text{LEL})} \tag{8.1}$$

with G as gas emission in g/h, f as quality factor of air flow distribution ($f = 1$ for even ventilation), T as exhaust air temperature in K, k_{adm} as safety factor (usually $k_{\text{adm}} = 0.5$, meaning that only 50% of the LEL are allowed), LEL as lower explosion level in g/m^3 at 20 °C (= 293 K) (BGR 104 2005).

Since the mass of sewage sludge in a bunker varies over time due to material discharge during operation, in practice, the air exchange rate is defined with the total bunker volume without considering the sludge bed (BGR 104 2005). In the VGB (originally "Vereinigung der Großkesselbesitzer", nowadays named VGB Power Tech e.V. located in Essen, acting as a technical association of operators of energy facilities) leaflet M 116 (VGB 1998), a minimum air exchange rate of 2 per hour is recommended for exhaust ventilation of generated methane. The bunker must be equipped with a methane warning system. Reaching a methane concentration of 20 or 40% of the LEL, depending on ATEX risk evaluation and eveness of exhaust ventilation, the air exchange rate is augmented to 6 per hour as a recommendation. Ventilation is provided by two ventilators that are monitored by pressure difference measurement. One is in operation, the other in stand-by mode.

Since methane rises due to its lower density in comparison to air, local exhaust ventilation of the air-methane mixture must be carried out in a high position in the headspace of the bunker. A computational fluid dynamics (CFD) simulation is recommended, since uneven headspace ventilation results in locally higher concentrations of methane and thus in a higher quality factor f and higher exhaust air flow.

References

BGR (Berufsgenossenschaftliche Regeln) 104 (2005) Explosionsschutz-Regeln. Regeln für das Vermeiden der Gefahren durch explosionsfähige Atmosphäre mit Beispielsammlung. Hauptverband der gewerblichen Berufsgenossenschaften Fachausschuss „Chemie". Carl Heymanns Verlag, Köln

DWA (2020) Specific values for valuation of product quality of boilogically stabilized sludge (In German). Arbeitsbericht der DWA-Arbeitsgruppe KEK 2.1 „Stabilisierung und Entseuchung". KA 67; 12: 962–969

Frechen FB (2004) Odor Emission Inventory of German Wastewater Treatment Plants – Odor Flow Rates and Odor Emission Capacity. Wat Sci Tech 50(4):139–146

Heindl A, Dobslaw D (2022) Methane in the Storage and Dehydration of Sewage Sludge. Int Special Ed Korrespondenz Abwasser Abfall, 20–27

Huning (2022) Photo of Steel Bunker of Huning Umwelttechnik GmbH & Co. KG, Melle

Saxlund (2022) Sliding Frames – Optimum Discharging Devices for round silos with flat floor. Enterprise

Saxlund (2022a) https://www.saxlundgroup.com/de/medien/fallstudien/sliding-frame-sludge-silo-farnham/. Accessed 15 Aug 2022

Saxlund (2022b) Photos delivered by Saxlund International Soltau

VGB Merkblatt M 116 (1998) Fire and Explosion Protection for Drying and Incineration of Sludge (In German). Technische Vereinigung der Großkraftwerksbetreiber e. V. Essen

Conveyance of Dewatered and Dried Sludge 9

9.1 Types of Conveyors for Dewatered Sludge

Sufficiently dewatered sludge coming from a screw press or a centrifuge has a crumbly structure. For short distances, coreless screw conveyors are usually used; their operation is trouble-free at low speeds and low degrees of filling, affording larger screw diameters. Long distances in two dimensions are frequently bridged by scraper chain conveyors of rigid design. Three-dimensional conveying is carried out with pumps—either eccentric screw pumps or piston pumps. It must be emphasized that increasing mechanical stress completely changes the structure of sludge. It turns to a pasty and sticky mass after undergoing high-force impact due to mechanical conveying. The conveying capacity increases, especially in the case of screw conveyors, and the volume filling degree decreases, especially in the case of pumps.

High-speed screw conveyors with small diameters are not suitable for trouble-free long-term operation. Larger diameters and a lower degree of filling need to be used. High compression and traction forces, exerted by piston pumping on dewatered sludge, result in sludge cell rupture and the formation of volatile acids. Therefore, volatile organic compound (VOC) transfer will increase during drying, thereby increasing the effort required to clean exhaust air.

The design of the conveying system for dewatered sludge depends on several issues:

- Necessary lining inside and outside the building—as short as possible to reduce mechanical force on the sludge or a pressure increase in pumping lines.
- Piping bends are designed with a radius to pipe diameter ratio of > 5–10 to reduce system friction, wear, and energy demand.

A. Heindl, *Belt Drying of Sludge*, https://doi.org/10.1007/978-3-662-66512-1_9

- Integration of a separator for tramp materials (stones, screws, plastic parts, etc.) to protect subsequent machines, for example, extruders; for materials > 8–20 mm in diameter, the typical electric energy demand for a two-roller-separator is around 1.0–1.1 kWh/m^3 of dewatered sludge with a dry residue (DR) ranging from 20 to 30%.
- Content of sand or struvite ($MgNH_4PO_4 \times 8H_2O$ crystals) in dry mass (DM) of dewatered sludge.

Table 9.1 lists the types of conveying systems and their specific features.

Table 9.1 Types of conveying systems for dewatered sludge

Type of conveying system	Length per unit	Mechanical stress on sludge	Liability to sticking of sludge	Susceptibility to wear	Liability to tramp material, maximum single dimension	Flexibility in lining
Screw conveyor, horizontal or inclined	Horizontal lining up to 50 m coreless	Moderate to high, depending on speed	High	Moderate	High, maximum 30–50 mm	Low
Screw conveyor, vertical	Vertical distance up to 15 m	High	High	High	High	Low
Scraper chain conveyor	Horizontal and inclined lining up to 100 m	Moderate	Moderate	High	Low	Moderate in plane conveying
Conveyor belt	Horizontal and inclined lining up to 100 m	Low	High	Low	Low	Moderate in plane conveying
Eccentric screw pump, maximum DR ≈ 35%	With added tube up to 25 m	Moderate to high, pressure up to 25 bar	Moderate to high	High	Maximum 50 mm	Very high
Piston pump Maximum DR ≈ 45%	With added tube several hundred of meter	High, pressure between 20 and up to more than 100 bar	Moderate to high	Moderate	Maximum 2/3 of tube diameter at pump discharging point of sludge	Very high

DR, dry residue

For a scraper chain conveyor for the horizontal transport of dewatered sludge, a specific electric energy demand of around 0.07 kWh/(m^3 of dewatered sludge × m of transport length) (Saxlund 2022) is indicated, when limed sludge of 25–35% DR is discharged.

A typical electric energy demand for horizontal screw conveying of dewatered sludge is around 0.05 kWh/(m^3 of dewatered sludge × m of transport length). Screw conveyors (auger conveyors) for dewatered sludge are usually equipped with coreless screws to avoid blocking. Vertical auger conveyors have a peripheral speed of around 0.9–1.1 m/s at the outer diameter of the screw blade, and their specific electric energy demand ranges from around 0.10 to 0.12 kWh/(t of dewatered sludge × m of transport height). Pressed feeding of the vertical augers is essential for their proper functioning. Sludge that has been treated with high mechanical stress increases the power demand of the vertical screw conveyor due to augmented sticking to blades.

Eccentric screw pumps always require proper, uninterrupted feeding of dewatered sludge to avoid premature wear of the stator due to half-dry operation. In the case of pumping of Bio-P sludge, chrome plating of the rotor is carried out to increase its lifetime.

Piston pumps operate at sludge transport speeds of around 0.05–0.30 m/s, resulting in a pressure loss of 0.21–0.27 bar per m of pumping length for a digested sludge at 30% DR and a tube diameter of 140 mm (Zey 1995). The moderate increase in pressure drop with speed is related to a water film that builds up between the sludge and the tube wall during transport. Increasing DR augments pressure loss per m of transport length, but other factors also influence pressure loss and consequently the power demand of piston pumps (Zey 1995):

- Tube diameter and consequently conveying speed for a given sludge mass flow
- Consistency of dewatered sludge
- Type of sludge stabilization
- Grain size distribution in sludge
- Volatile dry residue (VDR)

Pumping is possible even at higher DR of dewatered sludge, as reported in 1990 for the Simmering waste incineration plant in Vienna, Austria, when highly dewatered sludge of 42% DR was pumped to fluidized bed incinerators. The pressure drop (headloss) was calculated to be around 1.75 bar/m without slip agent and half the value with slip agent (0.1–1% of sludge mass flow, diluted by a factor of 1:50 with water [Zey 1995]) for a pipeline of around 40 m length, a diameter of 140 mm, and a transport speed of around 0.053 m/s (NN 1990).

For sludge with 20% DR, the pressure drop in psi/ft is determined by (WEF 2010):

$$\Delta p = \frac{15.68}{D^{1.28}} + 0.245 \times \frac{Q}{D^2} \tag{9.1}$$

Fig. 9.1 Two eccentric screw pumps feeding two belt dryer lines equipped with pressure control. (HUBER 2021)

with D as inner tube diameter in inches, and Q the flow of dewatered sludge in gallons per minute (GPM). This formula is valid for tube diameters of 100–300 mm and sludge velocities ranging from 0.015 to 0.43 m/s (0.05–1.4 ft/s).

An extreme example of pumping dewatered sludge with 28% DR (1060 kg/m^3) via a 150-m long pipe of 305 mm inner diameter results in a headloss of 2933 kPa or 29.33 bar or 0.195 bar/m. This very low value is related to the low sludge speed of 0.104 m/s due to the large tube diameter of 305 mm (WEF 2010).

Struvite occurs in digested dewatered sludge at wastewater treatment plants operating with biological phosphate elimination. From practical experience, it is known that with a struvite content of more than 1 mass% of DM in dewatered sludge, wear in pumps for dewatered sludge significantly increases. Rotors of pumps and sealing elements are particularly affected. Wear might be reduced by separating a portion of the struvite out of thin sludge before dewatering or by choosing special wear-resistant materials.

Fig. 9.1 shows two eccentric screw pumps, each with a small buffer bunker of around 10 m^3 and feeding extruders of two belt drying lines (HUBER 2021). For long-term operation, eccentric screw pumps should have larger diameters and lower speed to reduce mechanical wear. The maximum pressure in pumping lines with eccentric screw pumps are normally kept below 16–20 bar absolute. The typical average speed of sludge is up to 2 m/min. DR contents exceeding 28% might require inner wall lubrication of pipes with water or flocculating agents to limit friction and maximum pressure in the pipes immediately behind the pump. Highly stabilized digested sludge or sludge treated with

Fig. 9.2 Vertical screw conveyor for dewatered sludge with distancing of around 8 m. (Heindl 2019)

hydrolysis might reach a DR content of 30–40%. During periods of maintenance, the residual amounts of dewatered sludge in the pipeline can be emptied by injecting compressed air. This prevents a pressure increase and the generation of odorous gases in pipes.

Fig. 9.2 illustrates a vertical screw conveyor feeding a horizontal screw conveyor with dewatered sludge.

9.2 Types of Conveyors for Dried Sludge

Dried sludge needs to be conveyed to a storing system for further use. The sand content (non-hydrochloric acid [HCl]-soluble content of residue of ignition) of municipal digested sludge—responsible for wear in mechanical or pneumatic conveying systems—commonly reaches a maximum of 13–15% of DM. However, values of more than 20% can be encountered, especially in the case of wastewater treatment plants with high sand loads of inlet wastewater and/or improper functioning of the sand separation system.

The design of the conveying system for dried sludge depends on several issues:

- Necessary lining inside and outside the building—as short as possible and as straight as possible
- Limited production of dust—due to higher susceptibility to self-ignition of dried material with higher dust content
- Wear of system (speed and volatile dry mass [VDM] of dried sludge)—system should operate at least 8000 h without maintenance, requiring lower speeds of conveying systems
- Admissible temperature of dried material before loading to a closed silo—usually less than 40–50 °C
- Continuous and discontinuous discharging of dried granules—design factor up to 2 for discontinuous discharging, usually as occurring at the discharge end of belt dryers
- Atmosphere explosive (ATEX) zoning depending on dust content of dried sludge and dust generation during conveying

Table 9.2 lists various conveying systems and their special features. Dosing of material must always be arranged as evenly as possible. In extreme cases of plate formation, breakers can be applied. Dried sludge tends to form bridges in conveying systems, and the system design of, for example, cellular sluices or connecting pieces needs to be designed with sufficiently dimensioned openings to ensure easy flow of the material. In some cases,

Table 9.2 Conveying systems for dried sludge

Type of conveying system	Length per unit	Mechanical stress on sludge	Liability to dust production	Liability to wear	Flexibility in lining	Electric energy consumption
Screw conveyor	Up to 6 m, usually with shaft	Moderate to high, depending on speed	Moderate	Moderate to high, depending on speed	Low	Low
Bucket elevator	Up to 15 m and more, easy transport upwards	Low	Low	Moderate, buckets are affected	Moderate in plane conveying	Very low
Pneumatic conveying system (continuous or discontinuous craft conveying)	Up to 100 m and more	High to very high	High	High to very high depending on number and type of bends	Very high	High

low-speed grinding machines are able to reduce grain size without increasing dust content to a higher extent.

The radius of bends of pneumatic conveying systems must be equal or more than eight times the pipe diameter, and special wear design must be foreseen. Air speed is around 20–25 m/s. The mass ratio of dried sludge in conveying air must be less than 4 kg per kg of air for continuous thin-stream conveying. ATEX must be considered inside the system and zoning performed accordingly. Adequate electrical grounding of the whole pneumatic conveying system is essential. For trouble-free operation of a pneumatic conveying system, the DR of dried sludge must be at least 85%, ideally over 90%. Heating-up of conveying air due to high compression of air and high resistance of the pneumatic conveying system needs to be avoided through proper dimensioning of pipes and reducing the line length and bends in the dryer hall. In extreme cases, it may be necessary to cool conveying air.

Cooling of dried sludge can be performed in a water-cooled screw conveyor or simply by drawing through ambient air in counter-current to material flow inside the conveying system.

The temperature of dried sludge when fed to a silo must be measured to avoid an increased risk of self-heating of the dried sludge during storage.

Fig. 9.3 illustrates a bucket elevator conveying dried sludge from two belt dryer lines (Heindl 2021). Fig. 9.4 shows a pneumatic conveying system that is fed by three screw conveyors (Heindl 2022).

Fig. 9.3 Bucket elevator system conveying dried sludge granules of two belt drying lines. (Heindl 2021)

Fig. 9.4 Pneumatic conveying system for dried sludge leaving a belt dryer; note the special design of bends. (Heindl 2022)

An estimation of the mechanical energy consumption P_m of bucket elevators in kW can be formulated by (WEF 2010):

$$P_m = Q \times H/103$$

with Q as throughput in ton of dried material/h and H as conveyance height in m.

References

Heindl (2019) Photo of a Vertical Screw Conveyor for Dewatered Sludge

Heindl A (2021) Photo of a Bucket Elevator for Conveying of Dried Sludge

Heindl A (2022) Photo of a Pneumatic Conveying System for Dried Sludge

HUBER (2021) Photo of a pair of eccentric Screw pumps Feeding a Pair of Belt Dryers

NN (1990) Feeding of Fluidized Bed Ovens with Dewatered Sludge (In German). Wasser Luft und Boden wlb 9: 70

Saxlund (2022) https://www.saxlundgroup.com/de/medien/fallstudien/sliding-frame-sludge-silo-farnham/. Accessed 15 Aug 2022

WEF (2010) Water Environment Federation Desin of Municipal Wastewater Treatment Plants, 5th edn. WEF Manual of Practice No. 8. McGraw-Hill, New York

Zey W (1995) Pumps for Dewatered Sludge – Design and Application (In German). Die Bibliothek der Technik 113. Verlag Moderne Industrie, Landsberg

Properties and Storage of Dried Sludge 10

There is a broad range of possible applications for fully dried sludge granulates:

- Fertilizer, for example, for golf courses and municipal greens
- Fuel for co-incineration plants
- Gasification and pyrolysis processes
- Fuel for cement production, delivering energy and minerals
- Fast control fuel for fluidized bed incineration of partially dried sludge to regulate temperature

Each application requires individual characteristics of dried sludge with regard to grain size, hygienic status, and dry residue (DR).

10.1 Characteristics of Belt-Dried Sludge

Typically, belt drying of sewage sludge is performed to a final DR content of more than 90%. Very often, full drying is described with a DR content of more than 85%, and designers start to plan with DR contents of less than 90%. However, it is strongly recommended to dry sewage sludge in belt dryers to a DR content of 90–94% for several reasons:

- Drying to lower DR always results in uneven drying, i.e., uneven moisture distribution in the bulk as a whole and in the individual granules, and increases the risk of self-heating during storage in vertical silos.
- Pelleting can be easily performed by dosing a small amount of water.
- Microbial stability is not sufficient as long as DR content does not exceed 87%.

The microbial stability of dried sludge depends on DR content, as illustrated in Fig. 10.1 (Heindl and Kurzweil 2018). At a DR content below 85–87%, microbial activity starts to

A. Heindl, *Belt Drying of Sludge*, https://doi.org/10.1007/978-3-662-66512-1_10

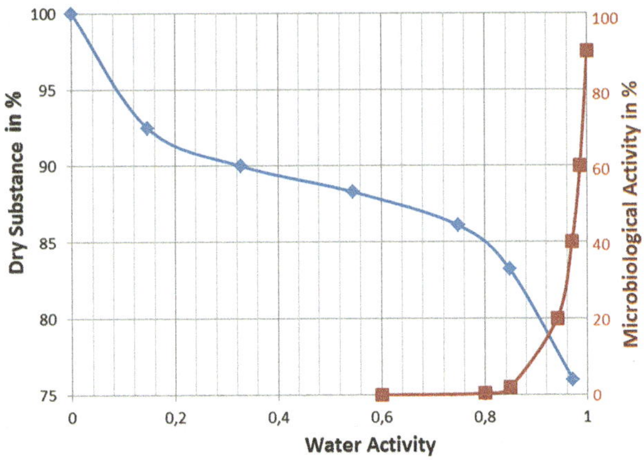

Fig. 10.1 Microbial activity versus availability of water expressed as water activity. (Heindl and Kurzweil 2018)

increase as the availability of water for metabolic action rises. Water activity describes the availability of water to microorganisms for metabolism and growth and corresponds to the relative humidity next to the surface of the dried sludge.

However, drying to a DR above 94% leads to higher dust formation as well as loss of dryer capacity and increases the risk of self-heating during storage in vertical silos.

Fig. 10.2 illustrates belt-dried sludge that has been extruded before drying.

Table 10.1 illustrates properties of belt-dried sludge with regard to bulk density, range of particle diameter, and portion of dust.

Different conditioning of dewatered sewage sludge entering the belt dryer influences the bulk density, porosity, and dust content of fully dried sludge.

In the case of back-mixing of a portion of dried sludge to obtain a sludge mixture of more than 60–65% DR, sludge granulates show a higher bulk density together with a lower inner porosity. Thus, on the one hand, shipping volume is reduced, but on other, inner

Fig. 10.2 Extruded and belt-dried sludge. (Heindl 2018)

Table 10.1 Characteristics of belt-dried sludge. (Modified from DWA M-379)

Belt drying system	Final dry residue	Bulk density in kg/m^3	Range of particle diameter in mm	Portion of dust, i.e., diameter of less than 500 μm
Extrusion	> 90%	350–500	0.1–10.0, average 5, maximum 20	< 1–2.5 mass% at dryer discharge
Back-mixing	> 90%	500–650	0.5–8.0	Dust is sieved and back-mixed

porosity and therefore the ability for water uptake in fertilizer applications is reduced. It is essential that dried material is sieved to minimize dust content and to recirculate mainly fines for mixing with dewatered sludge. The bulk density of dried granulates reaches 500–650 kg/m^3 and grain size covers a range of 0.5–8 mm at the dryer exit. However, dried material must be sieved, and thus a narrower grain diameter spectrum is adjusted.

Sludge formed by extrusion of dewatered sludge results in granulates of higher inner porosity and consequently of lower bulk density. No sieving of dried material is necessary, and dust content is usually less than 1.0–2.5 mass% when material leaves the belt dryer. Bulk density ranges from 350 to 500 kg/m^3 and grain size from 0.1 to 10 mm with maximum values of up to 20 mm. Oversized material is broken down into smaller pieces during mechanical conveying following dryer discharge.

Table 10.2 lists the grain size distribution of fully dried granulates that have been prepared by extrusion in front of belt drying (HUBER 2010–2021). The first three columns cover the analysis of dried sludge immediately after the dryer, whereas column four and five show values for dried sludge after a crusher and after pneumatic conveying.

Pneumatic conveying systems might cause higher dust contents in dried granulates of more than 10 mass%, thus causing a far higher liability to self-ignition of dried material in closed silos. Therefore, bucket elevators, which exert far less mechanical stress on

Table 10.2 Grain size distribution of fully dried sludge granulates formed by extruding before drying. (HUBER 2010–2021). Typical extrusion diameter of 10/13 mm round hole (*Note: data in columns refers to measurements of different sludges*)

Grain size in mm	1 after dryer (%)	2 after dryer (%)	3 after dryer (%)	4 after dryer and crusher (%)	5 after pneumatic conveying (%)
> 10	61.8	13.4	1.5	0.2	31.5
4–10	33.0	64.1	71.9	66.9	
2–4	3.1	17.2	20.5	28.0	
1–2	1.0	3.7	3.1	3.0	42.4
0.50–1.00	0.4	0.6	1.1	0.5	14.4
0.25–0.50	0.2	0.3	0.6	0.3	5.0
0.10–0.25	0.2	0.3	0.6	0.2	6.6
0.05–0.10	0.2	0.3	0.5	1.0	
< 0.05	0.1	0.2	0.3	0.1	
Dust content	≈ 1.1	≈ 1.6	≈ 1.9	≈ 1.6	≈ 11.6

Fig. 10.3 A crusher for fully dried granulates. (Heindl 2021)

sludge granulates, are preferred as long as the local situation allows the linear and vertical transport of dried material to the silo.

Granulate diameter can be reduced by a slowly moving crusher to, for example, less than 4 mm, thus increasing bulk density. The peripheral speed is less than 1 m/s to reduce the generation of fines and to avoid the risk of sparking in the case of contact with metal (atmosphere explosive, ATEX). Fig. 10.3 shows a slow-moving crusher of this kind (Heindl 2021). Further mechanical or pneumatic conveying increases the content of fines and dust.

10.2 Storage of Fully Dried Sludge in Silos

Sludge temperature at the entrance of the silo significantly influences the time period of subsequent discharging with regard to self-heating of fully dried sludge. When being discharged from a belt dryer, the temperature of fully dried sludge varies considerably between 40 and 80 °C, depending on:

- Air temperature of the dryer
- Existence of a cooling area for dried sludge in the last section of the dryer before discharge
- Efficiency of natural cooling of dried sludge by radiation and conduction during sludge transport to the silo

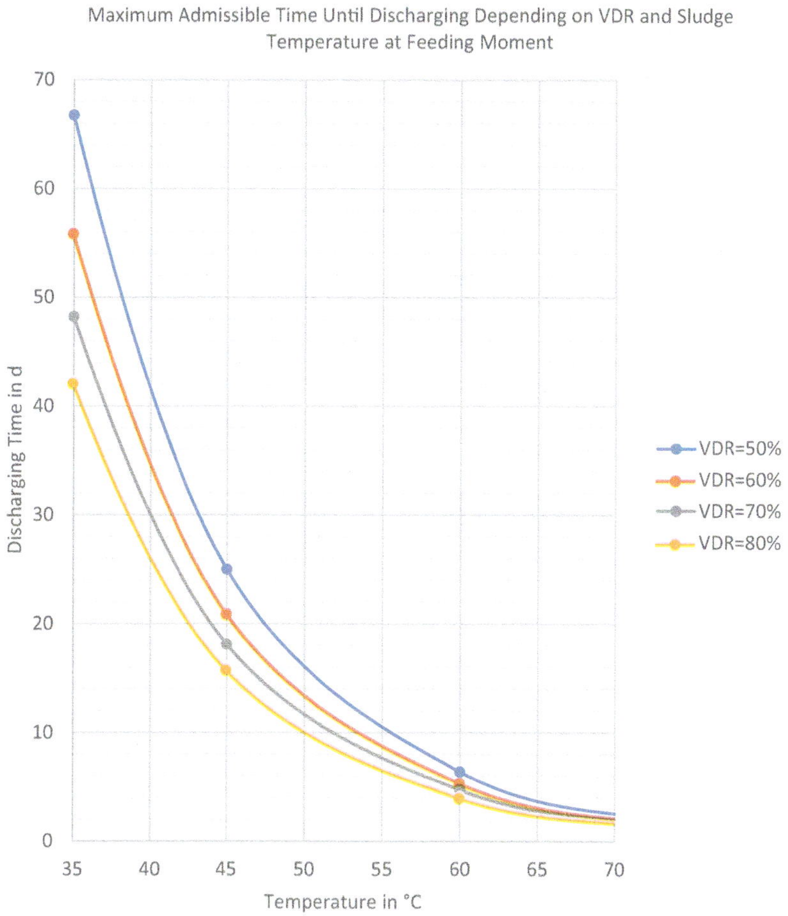

Fig. 10.4 Maximum time to discharge of a silo filled with fully dried sludge. *VDR*, volatile dry residue

Therefore, a sludge cooling system, either air- or water-cooled, needs to be arranged in the sludge conveying line following the dryer as soon as higher sludge temperatures above 50 °C occur.

Fig. 10.4 clearly illustrates that, based on theoretical calculations, higher sludge temperatures reduce the maximum allowable time between discharging actions for safe storage in silos. The influence of volatile dry residue (VDR) diminishes with increasing temperature of the sludge when fed to a silo. At 60 °C, the theoretical calculation requires that the silo be discharged every 4–7 days. A sludge temperature of 50 °C results in discharging periods of less than 12–18 days. The influence of the silo diameter is not considered, since the thermal conductivity of a bed of dried sludge granules is comparatively low, and even in silos with smaller diameters, the insulation capacity of bulk

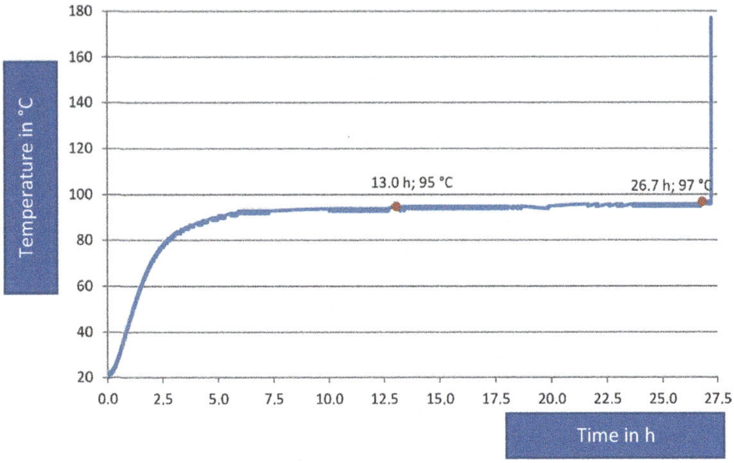

Fig. 10.5 Auto-ignition of dried sludge heated to 95 °C in an aluminum box. (Heindl and Kurzweil 2018)

material for conductive and convective heat transfer to the silo wall is relatively high. In any event, sludge temperatures of less than 50 °C are recommended nowadays, and in special projects, even less than 40 °C are required for safe storage. These low temperatures afford cooling of dried sludge where natural cooling on the transport route from dryer to silo is insufficient. Water- or air-cooled screw conveyors can be used.

A test with heating of fully dried sewage sludge taken from a municipal wastewater treatment plant with digestion was carried out at the Amberg-Weiden East Bavarian Technical University (OTH), Germany (Prof. Kurzweil). Sludge was placed in an aluminum box equipped with a thermocouple to measure sludge temperature and heated to 95 °C in an electric oven (Heindl and Kurzweil 2018). Fig. 10.5 demonstrates the temperature curve, whereby heating to 95 °C was completed in 13 h. Up to 26.7 h, only a small increase in temperature to 97 °C was observed. Then, however, a dramatic temperature rise could be determined, illustrating the exothermic energy released by the elimination of crystalline water from vivianite [$Fe_3(H_3PO_4)_2 \times 8H_2O$ or blue iron earth], overwhelming energy loss to surroundings.

The various possible reaction processes for self-heating of dried sludge that may occur during the storage of dried sludge are described in Fig. 10.6 (Heindl and Kurzweil 2018).

At ambient temperature, oxidation of FeS and FeS_2—boosted by water—can contribute to a temperature rise. Dried sludge with high iron contents (> 5–7 mass% in dry mass [DM]) in combination with high sulfur contents (> 1.5 mass% in DM) are susceptible to initial heating through exothermic reactions. Microbiological activity, which occurs at lower DR contents below 85%, causes a degradation of volatile mass and may also result in a temperature rise up to 80 °C. At sludge temperatures above 75 °C, vivianite is decomposed and crystalline water released. Water is adsorbed in the dried surrounding

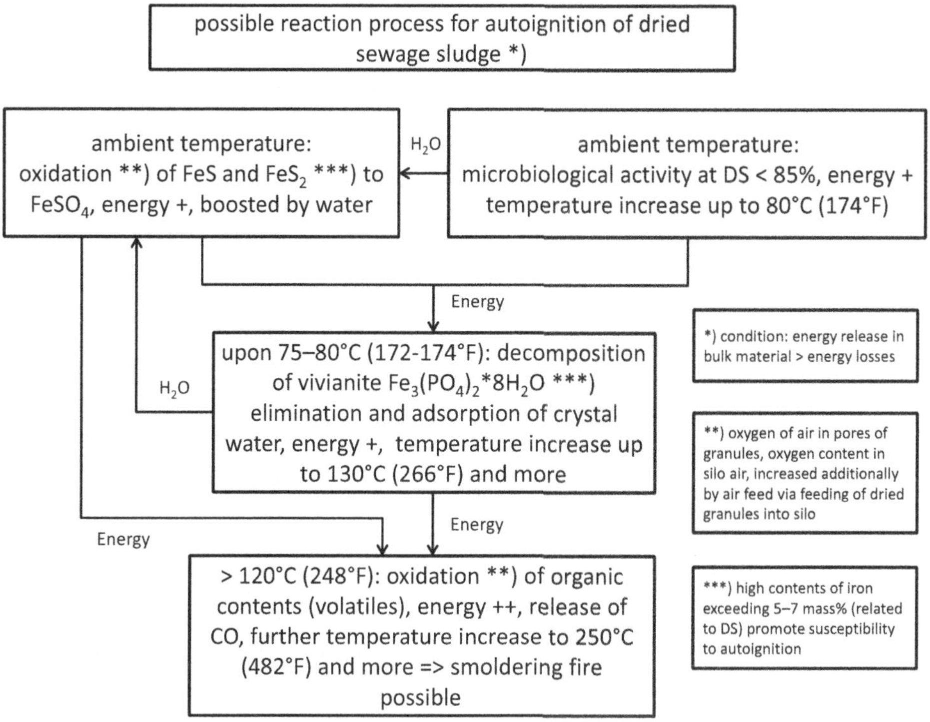

Fig. 10.6 Possible reaction process for self-heating of dried sludge during storage

sludge granules and an adsorption energy of 286 kJ/mole of adsorbed water generated. A further heating of sludge to temperatures above 120 °C occurs, and oxidation of volatile matter begins beyond temperatures of 80 °C (Poffet 2007), especially for dust particles with a very high surface area—relative to their volume—for oxygen adsorption. Carbon monoxide (CO) is generated and further heating of the sludge mass initiated until the temperature exceeds 220–250 °C and self-ignition occurs. In practice, glowing spots in stored dried sludge were surrounded by wetted granule agglomerations, confirming the self-heating effects explained here.

In the past, several issues with smoldering of dried sludge have been observed. CO values of 500–1000 ppm in the headspace of the silo can be encountered under normal operational conditions depending on sludge quality and ambient CO concentration. An automatic alarm is initiated for a CO increasing speed of more than 100 ppm/h (1.66 ppm CO/min). The average value of the last 30 min is compared to the actual CO concentration. Values rapidly exceeding 1600 ppm indicate smoldering inside the silo and inertization must be initiated (VGB 1998). In this case, at least five times the total silo volume needs to be inserted as N_2 to extinguish the smoldering fire in the silo. The residual concentration of oxygen in the silo must not exceed 4 vol%.

Fig. 10.7 Two silos for storing dried sludge at the special-purpose association for thermal sewage sludge utilization (ZTKS) Schwandorf, Germany. (Heindl 2022)

The standard safety equipment for silos comprises temperature measurement, CO and O_2 measurement, and inlet valves for inserting N_2 in the conical and cylindrical part of the silo.

The silos should be as gas-tight as possible in order to:

- Prevent oxygen from being drawn in for the case that a smoldering spot develops
- Prevent nitrogen losses during inertization

Fig. 10.7 shows a double silo for storing dried sludge, each silo having around 100 m^3 for storing two belt drying lines, each producing around 780 kg of dried sludge with 90% DR (Heindl 2022). In the foreground, two bundles of nitrogen pressure bottles can be seen for emergency use. The storage unit is additionally equipped with a stationary nitrogen production unit. Trucks can cross underneath the silo for easy and dust-proof discharging of the silo into a silo truck.

Lucas reported an incident of a smoldering fire with a CO gas explosion in the case of a 1000-m^3 silo for fully dried sludge (Lucas 2008). A drying plant was in the process of being commissioned and the silo was filled with 70 m^3 of sludge within 1 week. The

Fig. 10.8 Damaged discharging screw conveyor of a silo for fully dried sludge. (Lucas 2008)

sludge temperature was measured to be around 32 °C when stored and DR content ranged from 92–96%. Around 3 weeks after the first filling of the silo, approximately 8 m^3 were discharged into a truck and the CO content in the silo rose from 30 to 60 ppm, but went back to 30 ppm. Aeration of the discharging screw was only in operation during the discharging action. Within 1 day, there was a sharp increase in CO to more than 1000 ppm. Staff checked the silo by opening a hatchway—since a false reading from the CO probe was assumed—but could not detect any smoldering inside the silo. Within another 4 h, a deflagration caused by the ignition of CO occurred, damaging the discharging screw of the silo (Fig. 10.8). In addition, the steel of the silo cone connected to the screw conveyor showed tempering colors, a sign that smoldering of dried sludge with a strong temperature increase occurred.

The silo was equipped with CO measurement, temperature probes, and connecting flanges for inertization. The reasons for this incident were attributed to:

- Possibly higher sludge temperatures during the commissioning period or improper temperature measurement point in the dried sludge
- Wet sludge spots fed to the silo during frequent on/off operation during the commissioning period and related microbial activity
- Generation of CO by the smoldering action of the dried sludge in the conical part of the silo
- Low cooling effect due to the ambient temperature of around 32 °C
- Over-drying of material and formation of dust, thereby significantly increasing oxygen binding to dry sludge
- Non-sealed valve separating the silo from the discharging screw and diffusion of CO into the screw conveyor

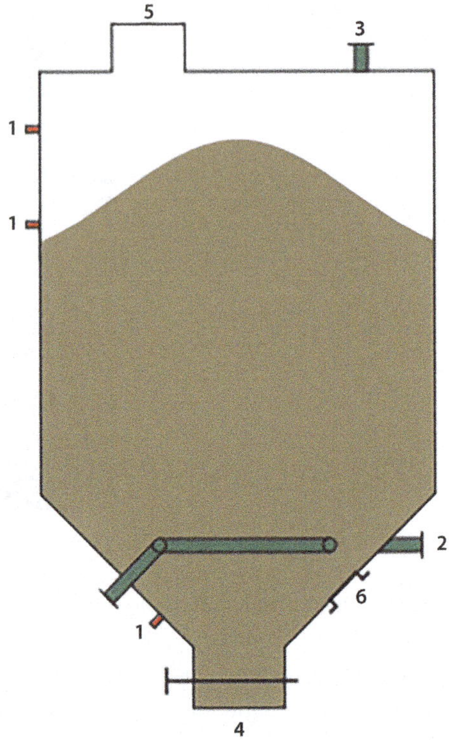

Fig. 10.9 Silo design with regard to safety issues. (VDS 2154 2008). *1* Additional openings for measuring probes 1/2 inch; *2* lower tube or flange for feeding of nitrogen at a height of 1–1.5 m above discharging; *3* upper inertization flange; *4* discharging; *5* feeding; *6* emergency discharging

This incident shows that, in the case of abnormal operating conditions, the risk for self-ignition increases and that the silo must be equipped with sufficient safety devices to avoid greater damage or even casualties.

In recent projects, total inertization with nitrogen (N_2) is carried out, thus preventing auto-ignition of dried sludge by reducing the oxygen content in the silo air to less than 4 vol%, as self-heating is still possible at 5 vol% (Lucas 2008). An additional measurement probe for oxygen must be installed.

Fig. 10.9 shows a recommended silo design with regard to safety issues (VDS 2008). The silo must be marked with the maximum admissible inner pressure.

Fig. 10.10 shows a huge silo for dried sludge granules with a tank for continuous inertization (HUBER US 2017).

Partial inertization is also practiced, whereby only the volume of dried granules is flooded with nitrogen. The energy demand is lower. Technical data are presented in Table 10.3.

A further table, Table 10.4, shows the technical data for fully continuous inertization of a 100-m^3 silo (total silo volume, see Fig. 10.10). One needs to accept a higher energy

Fig. 10.10 Total continuous inertization of a silo for dried sludge on the *left* side with a nitrogen tank to the *right* side for a belt-drying project in the Southeastern United States

demand in this case, and two additional bundles of nitrogen bottles are recommended for emergency cases. Inertization units of this kind comprise an air compressor, a dryer, a filter package, an air tank, a pressure swing adsorption (PSA) generator, and a nitrogen tank (Yara 2019).

Table 10.3 Technical data of a nitrogen producing unit for the continuous inertization of the bed of dried sludge in a 100-m^3 silo. (Bissinger 2019)

Nitrogen demand	3–6 m^3/h, maximum 25 m^3/h ($\approx$ 4 h)
Nitrogen concentration in the bed	$\approx$ 96 vol%, i.e., oxygen is $\approx$ 4 vol%
Nitrogen concentration in the headspace above the bed of granules	$\approx$ 90–95 vol%, i.e., oxygenis $\approx$ 5–10 vol%
Electric energy demand	$\approx$ 1.2 kWh/h, max. 6.0 kWh/h ($\approx$ 4 h)
Specific electric energy demand for continuous inertization by generating nitrogen out of compressed air	$\approx$ 12 Wh/m^3 of bed of sludge granules
Average nitrogen demand with regard to full silo volume	30–60 L N2/(m^3 silo $\times$ h)

Table 10.4 Technical data for the continuous inertization of the whole silo volume (bed of sludge granules and silo headspace). (Yara 2019)

Nitrogen demand	25.6 m^3/h
Nitrogen concentration in bed	$\approx$ 97 vol%, i.e., oxygen is $\approx$ 3 vol%
Nitrogen concentration in head space above bed of granules	$\approx$ 97, i.e., oxygen is $\approx$ 3 vol%
Electric energy demand	$\approx$ 6–8 kWh/h

The following recommendations can be made for the safe storage of dried sewage sludge in silos and piles (modified from Heindl and Kurzweil 2018):

- Prevention of thermal conditioning of the sludge to above 75 °C for several hours, for example, due to a malfunction of a conveyor system. This must be ensured by a non-bypassable automatic dryer control.
- Dry sludge temperature before storage: ≤ 40–50 °C, controlled in front of the entrance to the silo. Depending on the drying system and temperature of the dry sludge at the dryer outlet, a sludge cooling system may be required.
- Minimization of the dust content of the dry sludge before storage by using conveying systems that are gentle on the material, such as bucket conveyors, and with low fall heights.
- Drying to more than 90% DR in closed tower silos. Drying to more than 94% DR must be avoided due to increased dust formation in the bulk of dried sludge.
- Partially dried sludge: storage in open dump bodies or CO-controlled flat silos at DR contents of > 65–70%. The lower the dry residue, the shorter the silo emptying intervals should be.
- Silo emptying: at least once per week, but ideally twice per week.
- Avoidance of moist spots and agglomerates through uniform drying.
- Avoidance of rewetting: storage in dry piles protected against ingress of rainwater or moisture.
- Insulation of the silo and temperature-limited heating of the silo cone (≤ 40 °C) to avoid condensation inside the silo.
- In the case of large amounts of sludge: distribution to several smaller silos < 100 m^3 instead of storage in one large silo.
- Provisions for the inertization of the silo with N_2 or CO_2. If necessary, continuous controlled inertization of large silos > 100 m^3. It must be noted that CO_2 has a far higher density in comparison to air and that the carbon dioxide, which is harmful to human beings, concentrates at the bottom of the silo and in the nearby area in the case of leakage flow.
- Monitoring by temperature measurement and CO measurement in the silo: taking into account the warning levels for CO measurement and the basic CO load through oxidation of unsaturated fatty acids in the sewage sludge. Monitoring O_2 in the silo is recommended since decreasing O_2 content in the silo is a clear sign of the acceleration of auto-oxidation of dried material (Poffet 2008).
- At plant sites processing external sludge: monitoring of the iron and sulfur contents of the delivered sludge for high values or alteration. High iron contents above 5–7 mass% and sulfur contents above 1.5–2.0 mass% in the DR increase the tendency of dry sludge to spontaneously combust. Feeding of dried sludge with pneumatic conveying increases oxygen content in pores of the material, thereby increasing the risk of self-heating during storage.

In summary, it is recommended that in the case of large silos for the storage of fully dried sludge beyond a volume of 100 m³, cooling of dried sludge to a temperature below 40 °C and full inertization of the silo is carried out to minimize the risk of fire or explosion incidents.

In the case of self-heating of dried sludge in a silo, the residual concentration of oxygen is brought down to less than 2 vol% by inerting. The necessary nitrogen feeding must be around 2.75-fold of the maximum silo volume—taking into account uneven distribution of oxygen in the silo—with a reserve of an additional 25%, and inerting time equal to nitrogen feeding time should not exceed 30–60 min (Yara 2019). The gas is released into the silo via a ring tube circulating around the cylinder and at each feeding point via a mushroom distribution device inside the silo.

10.3 Fertilizer Application and the EPA 503 Standard for Hygienization

For fertilizer application, the hygienic status must be observed. The hygienic status of sludge is divided into classes A and B according to the EPA 503 regulation of the US Environmental Protection Agency. Table 10.5 presents the types of application and various restrictions for meeting quality requirements of the classes A and B according to US EPA 503. Please note that many countries in South America and the Middle East accept

Table 10.5 Definition of sludge classes A and B according to EPA 503. (Buchauer 2007)

Sludge class	A	B
Application	As fertilizer in private gardens, gardening shops, golf courses, green areas without any restrictions	As fertilizer on agricultural areas with restrictions
Limitations/ restrictions	Either fecal coliform counts < 1000/g or *Salmonella* sp. < 3 per 4 g of TS ***And*** Fulfill one of the following criteria: Sufficient specified thermal treatment Sufficient specified treatment to increase pH level, e.g., by adding lime Sludge with enterovirus < 1 per 4 g TS or < 1 worm egg per 4 g TS Suitable stipulated PFRP Suitable process equivalent to PFRP and confirmed by local authorities	Fecal coliform counts $< 2 \times 10^6$ per g ***Or*** Suitable PSRP ***Or*** Equivalent process confirmed by local authorities

PSRP, process to significantly reduce pathogens; *PFRP*, process to further reduce pathogens; *TS*, total solids

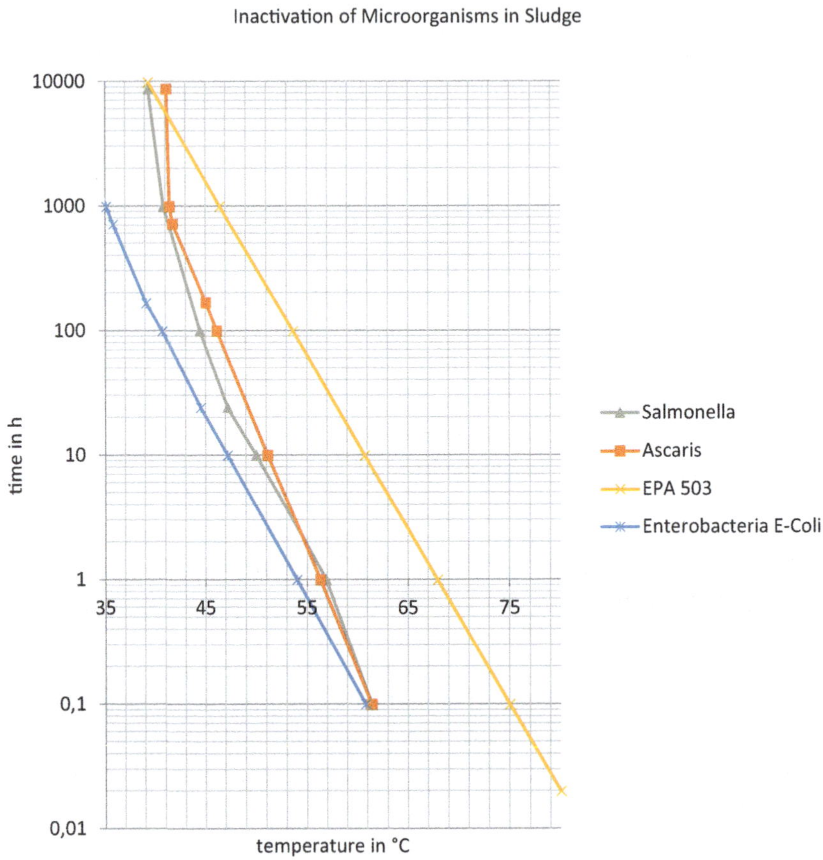

Fig. 10.11 Inactivation time (exposure time) of individual microorganisms depending on sludge temperature. (According to data from Feachem 1983)

EPA 503 regulations, but there might be local deviations from the US standard. The US publication "A plain English guide to the EPA 503 Biosolids Rule" EPA/832/R-93/003 is recommended for further study.

Fig. 10.11 shows the inactivation times of typical pathogens in sludge depending on temperature load (Feachem 1983). The EPA 503 inactivation line is shown in yellow. At a sludge temperature of 80 °C, the necessary time of inactivation is 0.02 h or 72 s.

Belt dryers are able to produce dried sludge meeting the Class A hygiene requirements of EPA 503. A common practice accepted by local authorities in the US is belt drying with a heat agent temperature of at least 95 °C, yielding an air temperature of at least 85 °C and a sludge temperature of more than 80 °C for at least 3 min. Sludge is dried to a DR content of more than 90%; in practice, greater than 92% is required for safety reasons to compensate water uptake of dried sludge during storing and distribution. The temperature–time–load of dried sludge must be measured and registered. This corresponds

Table 10.6 Results of chemical and microbiological analysis of dried sludge granules meeting EPA 503 regulations for Class A

Type of sludge	Mixture of waste activated and primary sludge, aerated for approximately 3–4 days
Temperature load on sludge	Belt dried with air temperature of around 85 °C
Dry residue	95.9%
Ammonium (NH_4–N)	1.900 mg/kg dry mass
Phosphorus (P)	12.000 mg/kg dry mass
Total Kjedahl nitrogen (TKN)	25.000 mg/kg dry mass
pH	6.2
Fecal coliforms	0.2 MPN/g dry mass
Concentrations of pesticides, metals, and organic compounds	Within fertilizer regulations

MPN,most probable number

to the alternative five biosolids treated in a process to further reduce pathogens (PFRP), i.e., drying to less than 10% water content and simultaneous heating to 80 °C (176 °F).

Table 10.6 reviews the results of chemical and microbiological analysis of dried sludge granules meeting EPA 503 regulations for Class A (HUBER US 2019).

10.4 Application of Dried Sludge in the Cement Industry

Using exhaust energy from cement production for sludge drying is a sustainable method for environmental protection. Cement clinker is ideally composed of around 68% CaO, 22% SiO_2, and 10% $Al_2O_3 + Fe_2O_3$. Dried municipal sludge has a similar composition in DM: 25–33% CaO, 38–48% SiO_2, 22–33% $Al_2O_3 + Fe_2O_3$. The brown ellipse in Fig. 10.12 represents the typical constituents of sludge in comparison to the gray circle of the cement clinker (according to Holpert and Schmidt 2016). Combining VDR with a high calorific value and its composition, belt dried sludge can be used as a mineral and energy source for the production of cement. The supply of minerals in dry sludge saves limestone and marlstone in cement production. The admixture amounts of dried sludge are limited to less than 10 mass%, especially due to the high phosphorus content of some sludges. The negative effects of specific sludge constituents on cement quality are presented in Table 10.7 (Holpert and Schmidt 2016).

Quality features are set by cement manufacturers, as shown in Table 10.8 as an example.

Another cement manufacturer limits the average content of mercury to less than 1 mg Hg/kg DM of dried sludge and rejects batches with more than 1.5 mg Hg/kg DM.

Waste energy sources of a cement factory include:

- Clinker cooler after electric filter: 200–350 °C
- Rotary kiln flue gas after electric filter: 300–350 °C

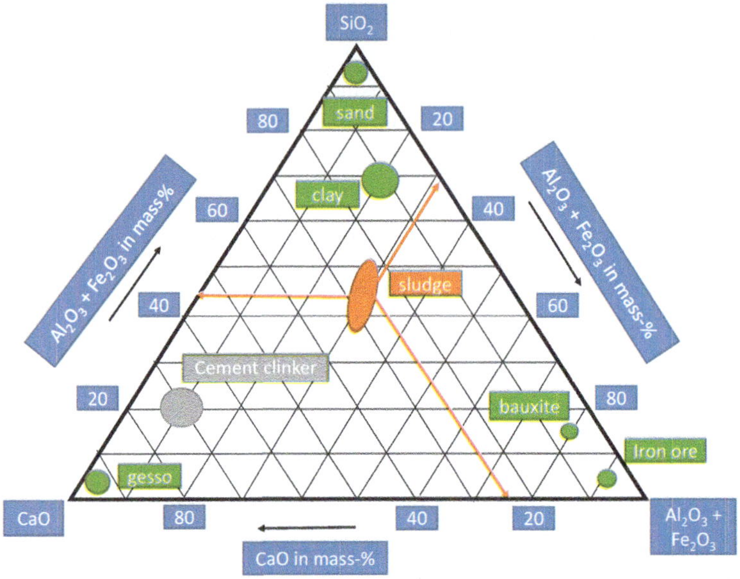

Fig. 10.12 Composition of sludge and cement clinker. (According to Holpert and Schmidt 2016)

In some cases, bag filter units are used to remove dust from clinker cooling air. However, their temperature limit of less than 200 °C forces the operators to add some ambient air to the clinker cooling air. Thus, the exhaust energy level is reduced and the energy balance negatively affected. Table 10.9 presents an evaluation of sludge drying at a cement site using exhaust air from clinker cooling as an energy source (Thamer 2011).

Due to the high temperature of clinker cooling air, it is possible to integrate an organic rankine cycle (ORC) for power generation of around 770 kW_e, assuming an electric efficiency of 15%. The temperature of thermal oil is reduced from 300 to 130 °C in the thermal oil circuit of the ORC, while the dryer receives thermal energy in the form of hot water with a feeding temperature of 110 °C in a second energy cycle. However, the dryer

Table 10.7 Negative effects of specific sludge constituents on cement quality, limiting feeding of sludge in clinker production. (Holpert and Schmidt 2016)

Substance/ ingredient	Normal concentration in cement in mass%	Negative effects of higher concentrations
Magnesia (Mg)	1.0–3.0	> 5.0 Increase in volume, self-destruction of cement products
Sulfates (SO_4^{2-})	3.0–4.0	Negatively affects solidification and strength
Alkali	0.5–1.2	Enhance early strength
Phosphates (PO_4^{3-})	0.2–0.6	> 1 Retards hardening
Manganese (Mn)	0.05–0.20	Affects cement color
Chlorides (Cl^-)	0.02–0.05	> 0.1 Corrosion of steel reinforcement

Table 10.8 An example of quality features set by a cement manufacturer

Quality feature	Limit
Dry residue in %	≥ 90
Lower calorific value in MJ/kg	> 8
Ash content in mass% of dry mass	< 60
Chloride in mass% of dry mass	< 0.2
Grain size in mm	Typically < 4 or < 10 in the case of pneumatic feeding into the rotary kiln for clinker production
Grain size distribution	For < 4 mm: 0–1 mm < 20 mass%, 1–4 mm > 80 mass%
Temperature in bulk when delivered in °C	< 45

Table 10.9 Evaluation of sludge drying at a cement factory using exhaust air from clinker cooling

Type of dryer	Single-belt dryer, air downwards
Type of sludge	From different sources with intermediate storage
Clinker cooling air and heat exchange	100,000 Nm^3 dry, 320 °C → 248 °C
Heat carrier thermal oil and heat exchange	50 t/h; 120 °C → 200 °C, heat loss 3%
Drying air temperature, initial	≈ 140 °C
Thermal energy demand of belt dryer	2625 kWh/h
Electric energy demand of belt dryer	230 kWh/h
Dryer exhaust air	≈ 9885 m^3 at 40 °C and 100% RH
Undigested sludge dried from 20 to 90% DR	930 kg/h
Dewatered sludge fed to belt dryer at 20% DR	4185 kg/h
Water evaporation	3255 kg/h
Specific thermal energy demand	0.81 kWh/kg water evaporation
Specific electric energy demand	0.071 kWh/kg water evaporation
Cooling water for direct condensing unit	87 m^3/h, 25 °C → 47 °C, 2.2 MW
Condensate and scrubber exhaust water	4.2 m^3/h at 47 °C

RH, relative humidity; *DR*, dry residue

needs more drying area to process the same quantity of sludge or to perform the same amount of water evaporation per hour (Thamer 2011).

As a rule, the following technical data for exhaust energy use are available for a clinker exhaust air temperature of 250 °C:

- Exhaust energy approximately 40–45 kWh/t of clinker production
- Clinker cooling air approximately 1000 Nm^3/t of clinker production
- Cooling of the clinker cooling air via finned tube heat exchangers with a temperature difference of approximately 80–120 °C

A typical mass ratio of clinker production in t/d to feeding of dewatered sludge in t/d with 25% DR is around 11 : 1 (Heindl 2016).

With regard to phosphorus recovery as stipulated in the German Regulation for Waste/Sludge of 2017, dried sludge can be used as long as the phosphorus content of sludge is less than 20 g P/kg DM. Therefore, phosphorus recovery from wastewater or sludge must be carried out at wastewater treatment plants.

10.5 Phosphorus Recovery from Ash: The Role of Belt-Dried Sludge

Fluidized bed incinerators normally use two points for the feeding of preheated incineration air. The primary air is used for sustaining the fluidized bed and fed from the bottom to the top of the incinerator. The secondary air is fed at the headspace of the incinerator, controlling temperature and consequently the emission of harmful gases from incineration. Belt-dried sludge can be fed pneumatically to the splash zone at the interface of the fluidized bed and headspace in the case of lower calorific value of the volatile dry mass (VDM) of sludge fed to the incinerator. Thus, dried sludge replaces primary energy carriers such as gas or oil as control fuel and additionally contributes to environmental protection. In addition, the phosphorus in the ash can be efficiently recovered and used to produce a full phosphorus fertilizer (by a rate of more than 80 mass% of phosphorus fed to the wastewater treatment plant), for example, by leaching it in dilute acid and subsequent spray granulation. Such a process uses around 0.3 kWh_e per kg of ash or around 0.15 kWh_e per kg of finished fertilizer (Phosgreen 2022).

A fluidized bed incineration plant coupled with a belt drying unit possesses very high flexibility in using sludge of different DR, i.e., partially dried or fully dried sludge. Thus, we obtain a partial flow of sludge mass to be dried to 90% in the belt dryer. This fully dried sludge is mixed with dewatered sludge to a DR content of around 39–45%, which is necessary for a thermally autogenous combustion. The mixing station for feeding a mixture of partially dried or fully dried sludge with dewatered sludge is arranged next to the incinerator. On the other hand, fully dried sludge from external sources can be used as control fuel when spread on the flash zone of the fluidized bed.

An example of an energy balance for phosphorus recovery from sludge ash and the resulting carbon dioxide footprint is presented. The calculation does not take into account CO_2 generation caused by the erection of buildings or the production of machinery. Data are related to the DM of dewatered sludge and the production of phosphorus. The carbon dioxide footprint is calculated on the basis of German carbon dioxide emissions based on mixed electricity generation in 2021.

Assumptions:

- Wastewater treatment plant with 75,000 P, German wastewater treatment plant size of 4B, which is obliged to recover phosphorus, without co-substrate for digestion, 100% external supply of electric current and 20% portion of primary energy carriers such as natural gas for the drying process

- Wastewater treatment with digestion, where sludge generation is rated to 15 kg DM/(P × a), corresponding to 1125 t DM/a or 4500 t of dewatered sludge/a with a DR of 25% or 1250 t of dried sludge with 90% DR
- Phosphorus content of 35 g P/kg DM

Annual mass of phosphorus: 1,125,000 kg DM/a × 0.035 kg P/kg DM = 39,375 kg P/a.

Ratio of recovery by fluidized bed incineration and processing via suspension in dilute acid with subsequent spray granulation: approximately 75% = 29,531 kg P_{rec}/a ≈ 30 t P_{rec}/a or ≈ 26 g P_{rec}/kg DM.

Belt drying of sludge: with electric energy consumption of 0.06 kWh_e/kg of water evaporation and 0.835 kWh_{th}/kg of water evaporation. For full drying from 25 to 90% DR per kg of DM, water evaporation of approxmately 0.722 kg is necessary. Specific energy demands will be 0.043 kWh_e/kg DM and 0.603 kWh_{th}/kg DM, resulting in 1.64 kWh_e/kg of P_{rec} and 22.97 kWh_{th}/kg of P_{rec}. Assuming drying with 80% of thermal energy covered by exhaust energy and total external purchase of necessary electric current: 366 g CO_2/kWh_e (D 2020) and portion of electric current 0.600 kg CO_2/kg P_{rec} as well as a portion of primary energy of 4.59 kWh_{th}/kg P_{rec} × 200 g CO_2/kWh_{th} = 0.918 kg CO_2/kg P_{rec}. Total: 1518 kg CO_2/kg P_{rec} caused by belt drying.

Transport: Fully dried sewage sludge is delivered to a mono-incineration plant: 1562.5 t/a with 90% DR → Transport volume with a bulk density of 400 kg/m^3 = 3906 m^3/a → 78 truckloads with 50 m^3 und 111 g CO_2/(t × km) for 100 km of transport distance (round trip) = 17,316 t CO_2/a or 0.577 kg CO_2/kg P_{rec} for transport.

Fluidized bed incineration of sewage sludge with in-house generation of electric energy and thermal self-sufficiency: 30,000 t DM/a, efficiency of generation of electric current of 7% relative to fuel power of 10.57 MW_{th}, 70% of electric current demand covered by generation of own electric current, 30% corresponding to 0.0592 kWh_e/kg DM of external electric energy supply, district heating of 6 MW_{th}. CO_2 footprint by external supply of electric current 2277 kg CO_2/kg P_{rec}, CO_2 saving by district heating (replacement of natural gas) for 7 months per year (6 MW_{th} × 8000 h/a × 0.583 = 27,984 MWh_{th}/a) × 200 kg CO_2/MWh_{th} = 5,596,800 kg CO_2/a or with 30,000 t DM/a approximately 7.175 kg CO_2/kg P_{rec}, CO_2 saving by applying fully dried sewage sludge as control fuel for incineration: assumed calorific value, approximately 11,000 kJ/kg of dried sludge or 12,200 kJ/kg DM or 3.39 kWh/kg DM or 0.678 kg CO_2/kg DM, yielding 26,077 kg CO_2/kg $P_{rec.}$

Production of fertilizer by spray granulation: A thermal energy demand of 120 kWh_{th} per t of granulate is assumed. The specific electric energy demand is set to 20 kWh_e per t of granulate. In addition, 1 kg of granulate corresponds to approximately 1.67 kg of DM, resulting in specific energy demands of 0.072 kWh_{th}/kg DM and 0.012 kWh_e/kg DM. In terms of phosphorus, this yields 2769 kWh_thh/kg P_{rec} and 0.462 kWh_e/kg P_{rec} as well as 0.554 kg CO_2/kg P_{rec} and 0.169 kg CO_2/kg P_{rec}.

Table 10.10 summarizes all assumptions and calculation steps for the energy balance and carbon dioxide footprint of phosphorus recovery from sludge ash.

Table 10.10 A summary of all assumptions and calculations for the energy balance and carbon dioxide footprint of phosphorus recovery from sludge ash

Phosphorus recovery from ash	Assumptions and calculations
Wastewater treatment plant	
Size	75,000 PE, German size class 4B with obligation to recover phosphorus
Wastewater treatment	Digestion, without co-substrate from external sources, external supply of electricity, 20% of thermal energy as natural gas for drying process
Balance for phosphorus	
Sludge accumulation	Approximately 15 kg DM/(P × a) or 1125 t DM/a or 4500 t dewatered sludge/(P × a) with 25% DR or 1250 t/a of dried sludge with 90% DR
Phosphorus content	35 g/kg DM
Annual amount of phosphorus	39,375 kg P/a
Degree of phosphorus recovery from ash	75% → 29,925 kg P_{rec}/a or approximately 26 g P_{rec}/kg DM and 0.399 kg P_{rec}/(P × a)
Belt drying	
Electric energy consumption	0.06 kWh_e/kg of water evaporation or 0.043 kWh/kg DM or 1.64 kWh_e/kg P_{rec}
Thermal energy consumption	0.835 kWh_{th}/kg of water evaporation or 0.603 kWh/kg DM or 22.97 kWh_{th}/kg P_{rec}
Assumption: external supply of electric energy, external supply of 20% of thermal energy demand as natural gas	Portion of electric energy 0.600 kg CO_2/kg P_{rec} with 0.366 kg CO_2/kWh_e (D 2020) and portion of natural gas 20% of thermal energy demand 4.59 kWh_{th}/kg P_{rec} with 0.200 kg CO_2/kWh_{th} (natural gas) → portion or primary energy 0.918 kg CO_2/kg P_{rec}
Transport	Delivery of fully dried sludge to plant with mono-incineration
Transport volume for bulk density of 400 kg/m^3	3906 m^3/a
Truckloads with 50 m^3	78
Transport distance round trip	100 km
CO_2 emission for transport with 111 g/(t × km)	17,316 t CO_2/a or 0.577 kg CO_2/kg P_{rec}
Fluidized bed incineration	Use of fully dried sludge as control fuel
Hours of operation	8000 h/a
Throughput of DM	30,000 t DM/a
Thermal fuel power	10.57 MW_{th}
Efficiency of power generation by steam turbine	7% Relative to fuel power, corresponding to 740 kW
Coverage ratio of demand of electric energy of incineration plant by steam turbine	70%

Table 10.10 (Countinued)

Phosphorus recovery from ash	Assumptions and calculations
External energy supply	30% = 0.0592 kWh_e/kg DM or 2.277 kg CO_2/kg P_{rec}
Energy supply for district heating (exhaust energy from condensing unit; 7 months/a) replacing natural gas	6 MW_{th} → 7.175 kg CO_2/kg P_{rec}
Use of fully dried granulate as control fuel for incineration	With calorific value of 12.200 kJ/kg DM: → 3.39 kWh/kg DM or 0.678 kg CO_2/kg DM or 26.077 kg CO_2/kg DM
Fertilizer production with spray granulation	
Thermal energy demand and CO_2 footprint	120 kWh/t granulate with 1 kg granulate = 1.67 kg DM; → 0.072 kWh_{th}/kg DM or 0.554 kg CO_2/kg P_{rec}
Electric energy demand and CO_2 footprint	20 kWh_e/t granulate → 0.012 kWh_e/kg DM or 0.169 kg CO_2/kg P_{rec}

Summary	CO_2 footprint due to electric energy consumption in kg CO_2/kg P_{rec}	CO_2 footprint due to thermal energy consumption in kg CO_2/kg P_{rec}
Drying	0.600	0.918
Transport	0	0.577
Fluidized bed incineration	2.277	−7.175 (district heating) ± 26.077 (control fuel) = −33.252
Of Fertilizer production	0.554	0.169
Total	3.431 kg CO_2/kg P_{rec}	−31.588 kg CO_2/kg P_{rec}
Total CO_2 footprint	−28.157 kg CO_2/kg P_{rec}	

PE, population equivalent; *DR*, dry residue; *DM*, dry mass

Thus, the use of fully dried sludge for the recovery of phosphorus from the ash of a fluidized bed incineration results—under the given assumptions—in a saving of CO_2 in relation to the phosphorus produced.

References

Bissinger (2019) Bissinger Behälter- und Stahlbau GmbH, 74374 Zaberfe, Germany. Info on Silo Inertisation. https://www.bissinger-gmbh.de/

Buchauer K (2007) Hygienic standards for using biosolids in agriculture. An international comparison (in German). DWA, Hennef

Feachem RG (1983) Sanitation and Disease – Health Aspects of Excreta and Wastewater Management. John Wiley & Sons, Chicester

Heindl A (2016) Dried sewage sludge as refuse derived fuel for cement works (Internal presentation HUBER SE)

Heindl A (2018) Photo of Fully Dried Sludge Granules. Dewatered Sludge was extruded before Drying

Heindl A (2021) Photo of a Crusher for Dried Sludge Granules

Heindl A (2022) Photo of a Double Silo for Fully Dried Sludge. Schwandorf

Heindl A, Kurzweil P (2018) Storage Behavior of Dried Sludge. Korrespondenz Abwasser Abfall 65(5):450–460 (In German, English translation available from author)

Holpert M, Schmidt E (2016) Future of co-incineration of sludge (In German). 49. Essener Tagung für Wasser- und Abfallwirtschaft, Essen, 2–4 March 2016

HUBER US (2017) Photo of a silo for dreid sludge with permanent inertization. Southeastern USA

HUBER US (2019) Analysis of a dried waste activated sludge with regard to EPA 503. Denver, NC 28037 USA

Lucas J (2008) Hazards, Phenomena and Protection Measures for Self-Ignition of Organic Bulk Materials With Emphasis on Dried Sludge (In German). VDI Berichte Nr. 2096. VDI-Verlag, Düsseldorf, pp 99–121

Phosgreen (2022) Download. https://phos4green.glatt.com/de/projekte/seraplant-haldensleben/. Accessed 08-06-2022

Poffet MS (2007) Thermal Runaway of Dried Sewage Sludge in the Storage Tanks. Doktorarbeit Universität Fribourg

Thamer D (2011) Dried Sludge as alternative fuel. Presentation Andritz

VDS 2154 (2008) Inertisation of silos in case of fire (In German). VDS Schadenverhütung GmbH. VDS 2154 2008-01(4) https://vds.de/fileadmin/Website_Content_Images/VdS_Publikationen/vds_2154_web.pdf download 29-08-2022

VGB Merkblatt M 116 (1998) „Brand- und Explosionsschutz beim Trocknen und Verbrennen von Klärschlamm“. Technische Vereinigung der Großkraftwerksbetreiber e. V. Essen

Yara (2019) Presentation Overall Explosion Protection Concept for Safe handling of Alternative and Biogenic Fuels – Inerting with CO_2/N_2. Germany, Bad Hönningen ((www.yara.com) and additional communication of details of operation)

11 Pelletization of Dried Sludge

To avoid dust generation during the handling of dried sludge, to improve storage behavior, and to reduce of transport volume, pelletization is a suitable option.

Typical pelletization units have either round cylindrical dies or horizontal flat dies. Throughput of pellet mills strongly depends on pellet diameter and typically ranges from 500 to 10,000 kg/h. The arrangement of a pellet press with a horizontal flat die is illustrated in Fig. 11.1 (HUBER US 2019).

Pelletization significantly increases bulk density. The bulk density of fully dried sewage sludge is around 350–500 kg/m^3 (21.9–31.2 lbs/ft^3) for drying extruded sludge strains and around 500–600 kg/m^3 (31.2–37.5 lbs/ft^3) for sludge dried with back-mixing. Fully dried sewage sludge formed to pellets has a bulk density in the range of 700–800 kg/m^3 or 40.8–49.9 lbs/ft^3. Typical pellet diameters are 3/4/6/8/10/13 mm, and pellet length ranges from twofold to fivefold of the diameter. Pellets of 3–4 mm diameter and reduced length are used as a long-term fertilizer, whereas larger diameters of 6–12 mm are used for thermal oxidation or gasification processes. A rough estimation of specific electric energy demand is around 50–70 kWh per metric ton of pellets, including the sieving and cooling step. The specific energy demand increases with decreasing pellet diameter.

In addition to lower volume demand for transport, there is the advantage of a lower surface to volume ratio for storage, thereby reducing the amount of oxygen adsorbed at the surface during storage and thus reducing the risk of self-ignition of stored dried material. Water uptake from ambient air is also far lower. This enables longer storage of pellets with minimum dust generation and the use of dried sludge as a seasonal fuel.

There is no need to add a binding agent as organic matter in sludge acts as an adhesive, and the sludge pellets obtain a brilliant surface and great hardness after forced cooling with ambient air.

The optimum water content of dried sludge to be formed into pellets is around 10–12% relative to total mass or a dry residue (DR) content of around 88–90%. In practical terms, sludge is dried to slightly exceed a DR of 90%, and water is added in the mixing screw, which also acts as a dosing device. It is of the utmost importance that the pellets, whose

A. Heindl, *Belt Drying of Sludge*, https://doi.org/10.1007/978-3-662-66512-1_11

Fig. 11.1 Pellet press for dried sludge with top-to-bottom feeding hopper, dosing and conditioning screw conveyor, and a horizontal flat die press of KAHL. (HUBER US 2019)

temperature has risen to more than 100 °C, are cooled down to 30 °C immediately after pelletization to obtain a stable and non-fissured pellet surface. Of course, the air leaving the cooler and having a sludge like odor has to be treated. During the cooling of heated sludge pellets, water is evaporated and the final DR reaches values > 90%. As a rule, the dust content of pellets is lower than 1 mass%. The highest circumferential speed is around 2.5 m/s and the highest sludge temperature does not exceed 120 °C (Kahl 2014). Horizontal die pelleting machines are able to process sewage sludge even with a DR content of down to 60%, but pellet hardness is lower. Special pellets with a length to diameter relation of around 1–1.2 can be produced and may be beneficial for use in common fertilizer spreaders.

Dried municipal sludge has a sand content of up to 13–15% relative to dry mass (DM; values for Central European digested sludge). Sand means the material residue after glowing at 550 °C and dissolution with HCl acid. Wear of dies and grinder rollers is considerable. The usual lifetime is around 800 h of operation for 4-mm pellets and around 3000 h for 10-mm pellets.

Fig. 11.2 shows the three stages of pellet processing from belt-dried sludge via conditioned sludge (conveying, water addition, and mixing) to pelletization, final cooling, and sieving for 4-mm pellets.

Fig. 11.2 The three stages of pelletization, from belt-dried sludge via conditioned sludge to pellets with a diameter of 4 mm. (HUBER SE 2018)

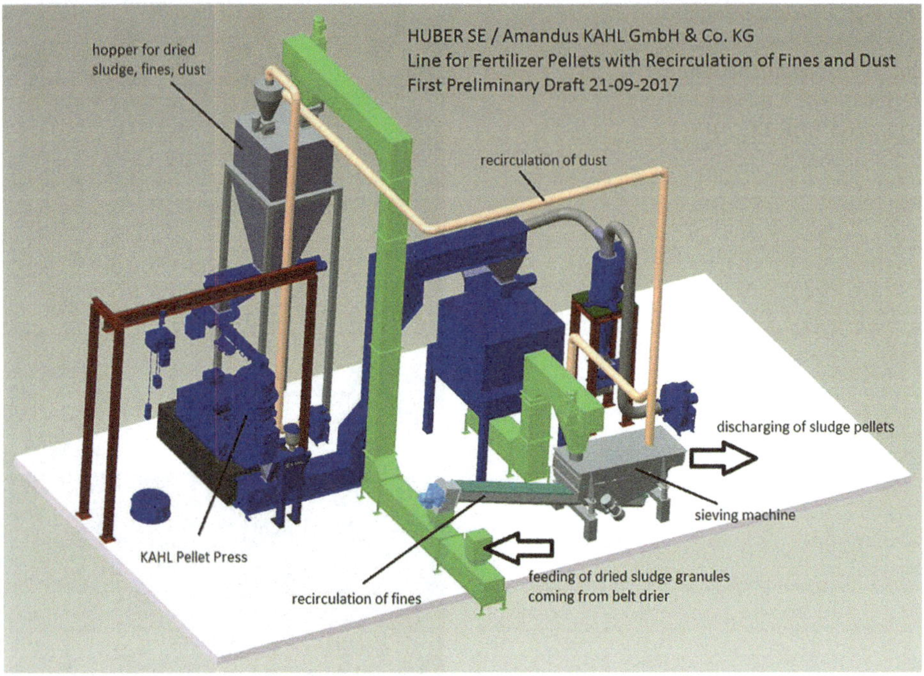

Fig. 11.3 Arrangement of a sludge pelletization unit with recirculation of fines. (Kahl 2017)

As dried sludge is (occasionally) milled, pressed to pellets, sieved, and cooled, dust generation inside some of the individual machines cannot be avoided. Therefore, dust explosion zoning must be performed for safety reasons. As a rule for European ATEX (ATmosphere EXplosive) regulations, Zone 21 is applied to the feeding, mixing, pelletization, cooling, and filtering units, whereas the storage silo/hopper for dried material is zoned as 20 and equipped with deflagration venting. Fig. 11.3 shows a typical arrangement of a pelletization unit with recirculation of fines (Kahl 2017), consisting of a hopper for dried sludge, fines, and dust, feeding and mixing screw conveyor, flat die pellet press, bucket elevator to cooler, counter-current cooler, bucket elevator to sieving machine, screw conveyor to recirculating bucket elevator, bucket elevator, and fan for cooling air with a dust separating cyclone.

For US conditions, the NFPA 820—as long as the dryer is installed on a wastewater treatment plant (standard for fire protection in wastewater treatment and collection facilities)—and the latest editions of articles 500/506 of the National Electrical Code (NEC) are taken for fire and explosion protection. Electrical parts are considered with regard to temperature class of dust, zoning, and electrical conductivity of dust. Typically, zoning or area classification is made with Class 2, Division 2, Group G inside and outside the processing line up to a distance of 10 ft around the machine assembly. This corresponds to Zone 22 according to the European ATEX. The storage hopper for dried material can be zoned inside with Class 2, Division 1, Group G, corresponding to ATEX Zone 21/20. Since 2017, the NEC also allows ATEX area classification.

In any event, area classification for a pellet-processing plant must be clarified in a timely manner and carried out by an expert to avoid problems in cost calculations and operating permits. Details are presented in Chap. 12.

References

HUBER SE (2018) Photos of dried granules and pellets of fully dried sludge. Berching

HUBER US (2019) Technical data and photo of a pellet press for fully dried sludge granules

Kahl (2014) Amandus Kahl GmbH & Co. KG Pelleting of municipal and industrial sewage sludge. Prospectus Digi-AK-41-E 04/14. Reinbeck/Hamburg

Kahl (2017) Amandus Kahl GmbH & Co. KG Arrangement of a Kahl pellet press with recirculation of fines. Reinbeck/Hamburg

Risks from Belt Drying Plants for Sewage Sludge 12

12.1 ATEX-Related Properties of Dewatered and Dried Sludge

During storage of dewatered sludge, further degradation of organic mass occurs, resulting in the generation of volatile organic compounds (VOCs), of which approximately 90 mass% expressed as Carbon is Methane (CH_4). Of course, this portion depends on the length of the storage period. To avoid the development of explosive gas atmospheres in bunker areas, the ventilation of bunker headspace is necessary. Details are presented in Chap. 8. As the mass of stored dewatered sludge varies temporarily during the operation of a drying plant, the air exchange rate is, in practice, related to the whole bunker volume without separate consideration of sludge bed volume (BGR 104 2000). The most important gases are presented in Table 12.1 with regard to their critical properties.

Due to its volatile dry residue (VDR) content, which typically ranges from 50 to 80%, dried sewage sludge is a solid fuel. Therefore, there are risks with regard to fire and dust explosions, meaning that safety equipment and measures must be installed and taken, respectively, to ensure safe operation of sludge drying.

Depending on the system for the preparation of dewatered sludge before belt drying, different contents of dust are found in the dried sludge. Therefore, when handling dried sludge,

Table 12.1 Critical properties of gases emitted during storage of dewatered sludge at atmospheric pressure. (Molnárné et al. 2008)

Properties	Gas type	
	Methane (CH_4)	Ammonia (NH_3)
Temperature class according to EN 600789	T1 (> 450 °C)	T1 (> 450 °C)
Temperature of ignition in air in °C	≈ 595	≈ 630
Lower explosion level in air in vol%	≈ 4.4 vol% (20 °C) ≈ 3.7 vol% (120 °C)	≈ 15.4 vol% (20 °C) ≈ 13.4 vol% (120 °C)
Lower explosion level in air in mg/m^3	≈ 29.4 g/m^3 (20 °C)	≈ 109.0 g/m^3 (20 °C)

A. Heindl, *Belt Drying of Sludge*, https://doi.org/10.1007/978-3-662-66512-1_12

dust will be present and the critical properties of sewage sludge dust must be observed to achieve safe operation of the drying plant. Whether or not zoning is necessary inside the belt dryer depends on the occurrence of dust contents. Dryers operating with extrusion of dewatered sludge for feeding normally need no zoning as the dust content is below 3 mg/m^3 of air leaving the dryer. In combination with an optimum dryer design, dust deposits will be small during longer periods of continuous operation. There is a high margin of safety to the lower explosion level for sludge dust ranging from 30 to 60 g/m^3 of air.

It must be emphasized that the critical properties of dried sludge change with temperature, particle size distribution of dust, and composition of the organic matter (fat, protein, or hydrocarbon content). Thus, when designing an adequate level of safety for a plant where only one or two kinds of sludge are processed, the critical properties of the sludge varieties should be measured.

A document published in the UK by the Health and Safety Executive (HSE) on the risks of sewage sludge drying plants deals with the details (HSE 2012). A summary of information on safety issues relating to dried sludge and the safe operation of sludge dryers can also be found in the DWA *Merkblatt-M 379* "Drying of Sewage Sludge" (2021, in German) and in Manchester (2001). In Table 12.2, values from the latter source are given in italics.

Ranges of critical properties of dried sewage sludge are summarized in Table 12.2 together with useful remarks.

Organic (sludge) dust exposed to temperatures above 120–150 °C for prolonged periods may transform into pyrophoric coal by a process of carbonization. This substance has a very high inner porosity and is able to take up oxygen to around 30- to 90-fold of its

Table 12.2 Critical properties of dried municipal sludge

Exploded dust	Explosion class (St)	1 (weak to medium explosion)
	Explosion constant (K_{St})	80–150 bar × m/s *80–200 bar × m/s*
	Maximum pressure of explosion (MPE)	6–9 bar *7–9 bar(g)*
	Minimum ignition energy (MIE)	1000 mJ > MIE > 100 mJ *250 mJ < MIE < 500 mJ* (125 mJ–750 mJ for inductive ignition)
	Minimum ignition temperature of a dust cloud (MIT) ATEX: T < 2/3 of MIT	360–510 °C *350–500 °C*
	Lower explosion limit (LEL)	60–500 g/Nm^3 *50–200 g/Nm^3*
	Limiting oxygen concentration (LOC)	6.5–12.5 vol% (100 °C) *9–12 (20 °C) for N_2 and CO_2, recommended safety value at least 4 vol% underneath*

Table 12.2 (Continued)

Deposited dust	Burning number (BN) (100 °C)	3 (local burning or smoldering without spread)
	Layer ignition temperature (LIT) of a 5-mm dust layer ATEX: T < GT°C–75 °C	220 (260)–340 (430)°C *150–280 °C*
	Self-ignition temperature (SIT)	400 cm^3 (D = H): 120–180 °C
	Spontaneous decomposition	None
	Impact sensibility	None
	Electrical conductivity	Group IIIB, non-conducting dust, specific resistance > $10^8 \Omega \cdot m$
	Marking of electrical components in ATEX area	Zone 22: II 3D T4 IP5x Zone 21: II 2D T4 IP6x

material volume, thereby reducing the layer ignition temperature to slightly above 150 °C (Müller-Hillebrand 1940).

It should be noted that hazard features of sewage sludge dust vary considerably and that for a specific sludge drying project, it is recommended to analyze the dust produced from the sewage sludge to be processed.

Table 12.3 presents experimental data on the critical properties measured for a dust of dried mixed sludge (PS + WAS; HUBER US 2017).

The dust layer depth was $\frac{1}{2}''$ and therefore more than twice the thickness for the test according to the European ATEX. Thus, it can be assumed that the layer ignition temperature (LIT) for a 5-mm dust layer will be higher than 240 °C.

Table 12.3 Critical data of a dust sample, generated from a mixture of PS and WAS

Critical property	Value
Particle size	Mean 28 µm, 93% < 75% of dust mass
Dry residue	98.8%
Explosion constant (K_{St}) (Siwek 20 L-chamber, ASTM E 1226)	213 bar × m/s ± 10%
Maximum pressure of explosion (MPE) (Siwek 20 L-chamber, ASTM E 1226)	8.2 bar ± 10%
Minimum ignition energy (MIE) (ASTM 2019)	≈ 14 mJ (without induction)
Layer ignition temperature (LIT) of a 12.7-mm dust layer (ASTM 2023)	240 °C

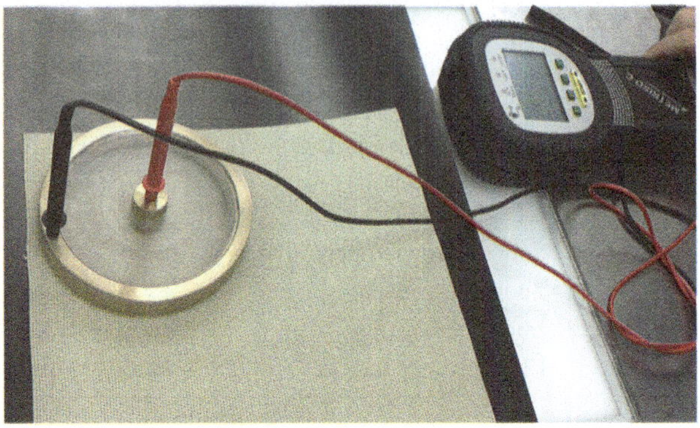

Fig. 12.1 Measurement of the electrical resistance of a polyethersulfone (PES) belt. (Heindl 2018)

12.2 Static Electricity as a Source of Ignition Caused by the Movement of Plastic Woven Belts

Measuring electric resistance of a polyethersulfone (PES) belt used for belt dryers resulted in 28.5 MΩ per 50 mm at ambient temperature (Fig. 12.1).

Resistance is greater than 1 MΩ. As a consequence, belts of this kind are liable to bunch discharges if electric charges are able to accumulate on the belt surface when moving along the dryer, especially at high belt speeds. Adequate grounding is necessary to avoid electric discharging. Normally, the belts are made endless by special stainless steel clamps running across the belt width, meaning that an electric connection to the belt-supporting rollers is periodically ensured and the accumulated electric charge released. Another option for releasing accumulated electric charges is to use woven belts with longitudinally integrated woven metal wires made of brass or stainless steel. For sludge drying, stainless steel wires are suitable.

12.3 Special Safety Recommendations for Directly Fired Dryers

Abbott highlighted the fire and explosion risks for belt dryers and recommended several safety measures for their safe operation (Abbott 1990). In any event, a distinction needs to be made between dryer systems operating with extrusion of dewatered sludge and systems applying partial recirculation of dried sludge to obtain a mixed sludge with a dry residue (DR) exceeding 60%. The key issue is dust development and dust deposits in the dryer combined with higher process temperatures. Dust development inside the belt dryer must be avoided by limiting air speed through the sludge layers on the belts. In this context, the extrusion of sludge is advantageous in comparison to back-mixing of sludge, as extruded strains form a compact and defined bed with a minimum of dust development. Belt dryers operating with back-mixing are often equipped with a dust measuring device to check that the dust content in drying air is far below the critical value of approximately 60 g/m^3 (26.2 gr/ft^3).

When using direct firing of the dryer, which is normally carried out for each section, Abbott recommends sectional vents spaced at a maximum of 6 m apart from each other and with an area of at least the vertical cross-section of the belt dryer. As fines might also agglomerate at the bottom of the dryer, the vents should be installed in side areas with ducting to a safe area outside. The vents must open fully at an internal overpressure of a maximum of 0.035 bar (3500 Pa or 0.50 psi), as belt dryers have large and relatively weak cross sections.

If the discharge section of the belt dryer is liable to dust generation, a dust hood should be installed and either explosion protection or venting can be used. According to the European ATEX regulations, the process temperature corresponding to the maximum temperature of the heating agent in the air heaters must be limited to 75 K below the minimum ignition temperature of a 5-mm dust layer of sewage sludge. Extensive analysis of sewage sludge dust showed that this minimum temperature is approximately 220 °C (428 °F; Lucas 2008).

12.4 Safety Integrity Level and Performance Level of Controls

12.4.1 Safety Integrity Level

The safety integrity level (SIL) means a safety standard for the control of a process according to IEC 62061. The SIL of a safety-related system stands for the risk reduction that the system must achieve. For continuously operating machines such as belt dryers for sludge, four SILs are specified (Table 12.4).

If a tolerable risk has a probability of 10^{-9} dangerous failures per hour and the equipment under control (EUC) has a probability of 10^{-2} dangerous failures per hour, then the protection system would need to have a probability of 10^{-7} dangerous failures per hour (Redmill 2000). The demands for the safety integrity of electronic and programmable electronic control systems need to be derived from the general risk assessment of the drying system. It is limited to risks coming directly from hazards caused by the dryer.

Some process steps are very important for the safety of belt drying of sewage sludge. Under unfavorable conditions, there may be a risk of fire or dust explosions. The tem-

Table 12.4 Safety integrity level: levels and probabilities of dangerous failure per hour

Safety integrity level (SIL)	Probability of dangerous failure per hour
4	$\geq 10^{-9}$–10^{-8}
3	$\geq 10^{-8}$–10^{-7}
2	$\geq 10^{-7}$–10^{-6}
1	$\geq 10^{-6}$–10^{-5}

perature of sludge and of drying air/heating agent plays an important role since dried or partially dried sewage sludge is liable to self-heating and self-ignition. Thus, for example, controlling the temperature of dried sludge leaving the dryer is very important for safe storage in a vertical closed silo. In the case of a product temperature exceeding 50 °C, dried sludge must not be conveyed to the silo, but instead cooled or stored in an emergency bin. The temperature control must be designed with two independent temperature measuring probes, each equipped with its own power supply. In belt drying projects, the measurement of dried sludge temperature was classified as SIL 2 in the past.

Of course, the SIL evaluation has to be carried out in cooperation between the operator and the supplier of the dryer. It must be borne in mind that higher safety requirements cause higher investment and control/maintenance costs.

12.4.2 Performance Level of Controls

The EN ISO 13849-1 defines features for safety controls, whose probabilities of failure are expressed by performance levels (PL) ranging from ‘a’ to ‘e.’ Table 12.5 presents the levels and the corresponding risks.

The following points need to be considered:

- Average probability of a dangerous failure per hour for each electronic component
- Common cause failure
- Behavior of safety function under failure conditions
- Systematic failures
- Ability to carry out a safety function under foreseeable conditions

Typically, the PL for belt drying systems is classified as level ‘c.’ Of course, the PL evaluation must also be carried out in cooperation between the operator and the supplier of the dryer.

Table 12.5 Levels of performance and corresponding risks

Performance level (PL) of control	Average probability of a dangerous failure per hour	
a	10^{-4}–10^{-5}	Low risk
b	10^{-5}–3×10^{-6}	
c	3×10^{-6}–10^{-6}	
d	10^{-6}–10^{-7}	
e	10^{-7}–10^{-8}	High risk

12.5 Hazard and Operability Study

A hazard and operability study (HAZOP), which comprises an analysis of risk, was originally set up in chemical industries. The HAZOP study simulates situations when process parameters encounter a deviation from the design intent. The impact on safety of incorrect operation by personnel or a failure of control or mechanical parts of the drying system is evaluated. The examination needs to prove that incidents of this kind do not endanger the lives of personnel or cause damage to the drying system. Also, emergency shut-downs and the impact on hazards are checked. Specifically, focus is placed on: temperatures exceeding maximum allowable temperatures in the dryer; issues related to dust or gas development during operation; microbial activity; or gas generation during standstill of a dryer filled with sludge. Issues relating to the storage of dewatered and dried sludge are also addressed (see Chap. 8 and 10).

HAZOP studies cover:

- Normal operation and operation deviating from the design
- Start–stop sequences
- Emergency shutdown
- Fire detection and extinction
- Prolonged standstill periods
- Failures in the supply of electric or thermal energy

References

Abbott J (1990) Prevention of Fires and Explosions in Dryers, 2nd edn. The Institution of Chemical Engineers, Rugby

ASTM (American Society for Testing and Materials) (2019) E2019-03(2019) Standard Test Method for Minimum Ignition Energy of a Dust Cloud in Air. West Conshohocken, PA, 19428-2959 USA.

ASTM (American Society for Testing and Materials) (2023) E2021-15 Standard Test Method for Hot-Surface Ignition Temperature of Dust Layers. West Conshohocken, PA, 19428-2959 USA

BGR 104 (2000) Berufsgenossenschaftliche Regel 104: Explosionsschutz-Regeln. Sammlung technischer Regeln für das Vermeiden von Gefahren durch explosionsfähige Atmosphäre. Jetzt DGUV-Regel 113-01

DWA (2021) Merkblatt M 379 Sludge Drying (In German). DWA (Deutsche Vereinigung für Wasserwirtschaft, Abwasser und Abfall e.V.), Hennef

Heindl A (2018) Photo of Measurement of Electric Resistance of a Woven Plastic Belt at GKD Works in Düren

HSE (2012) Health and Safety Executive Information Document (ID) on risks from Sewage Sludge Drying Plants. https://www.hse.gov.uk/

HUBER (2017) Critical properties of sludge dust of a project. HUBER Technology, Denver

Lucas J (2008) Hazards, Phenomena and Protection Measures for self-ignition of organic bulk materials, Focus on Dried Sludge (In German). VDI Berichte Nr. 2096, pp 99–121

Manchester SJ (2001) The Fire and Explosion Hazards of Dried Sewage Sludge. Symposium Series No. 148. IChemE.

Molnárné M, Schendler Th, Schröder V (2008) Safety Characteristic Data Volume 2: Explosion Regions of Gas Mixtures, 2nd edn. Wirtschaftsverlag NW, Bremerhaven

Müller-Hillebrand D (1940) Basics of the Installation of Electric Plants in Works Liable to Exposions (In German). Springer, Berlin

Redmill F (2000) Understanding the use, misuse and Abuse of Safety Integrity Levels. Proceedings of the 8th Safety-critical Systems Symposium, Southampton, 8–10th of February 2000 Springer

Important Points for Successful Mounting, Commissioning and Operation of Belt Dryers 13

13.1 Mounting and Commissioning of Belt Dryers

A well designed and executed concrete foundation as well as professional mounting of dryer elements are essential for the even running of belts. Maximum admissible deviations of measurements in longitudinal and transverse directions must to be observed to avoid side running of belts, even in the case of a well-adjusted tensioning of the belts during commissioning or of an installed belt guiding device.

Personnel must be well-trained on the job to ensure sufficient skills for maintenance and operation. Replacing skilled personnel with untrained workers often results in a decline in the quality of operation and a remarkable reduction in throughput. However, operation can be facilitated by a high degree of automation and a self-controlling dryer system. Measuring probes for the dry residue (DR) content of fed and discharged sludge, thereby providing a proper basis for changing dryer throughput to keep water evaporation of the dryer at a constant level, is a key issue in this context.

The procedure of commissioning and performance testing must be agreed between buyer/operator and seller of the belt drying unit. The duration of testing and maximum downtime need to be specified, as does the procedure of DR measurement or the admissible deviation from the final DR of dried sludge. The operator/buyer must take care of procuring a sufficient quantity of dewatered sludge with the agreed quality level and DR (±, e.g., 1% absolute of DR) and must ensure an energy supply (thermal/electric) meeting the stipulated quality and quantity. In some cases, the bulk density and dust content of dried sludge needs to be within certain limits to specify the necessary transport volume and to meet special specifications. All these figures and admissible deviations need to be agreed upon in advance. In the case of divergent conditions for performance testing, a conversion from actual to stipulated conditions needs to be agreed upon.

It is recommended that a definition of a reference sludge be agreed between the parties. Sludge features such as DR, volatile dry residue (VDR), and transfer of volatile organic compounds (VOC) in terms of mg C per kg of dry mass (DM) throughput of the dryer are

A. Heindl, *Belt Drying of Sludge*, https://doi.org/10.1007/978-3-662-66512-1_13

specified together with admissible deviations. The ammonium content and volatile sulfur content of dewatered sludge may be stipulated together with a range of admissible pH-values. Finally, not only the degree of stabilization but also the intermediate storage period from dewatering to drying can be specified. In this context, it is highly advantageous for the management of the drying plant to have an influence on the wastewater treatment of individual local plants, for example, in collaboration for sludge drying. In any event, lower disposal costs for sludge that is very well stabilized and rapidly dried will facilitate the operation of a sludge drying plant and reduce the costs of drying, especially with regard to exhaust air treatment. The definition of a reference sludge makes sense in the case of drying sludge coming from numerous different sources. Of course, specific sludge features strongly depend on the wastewater treatment process, degree of stabilization, and portion of industrial wastewater, and sludge handling influences the specification of a reference sludge.

13.2 Continuous Operation

Belt dryers are continuous drying systems designed to run 24/7. Only continuous operation guarantees even drying results and low specific energy consumption. Operating belt dryers for 8 or 10 hours per day results in frequent shut-down processes with negative effects on the economics of operation and may also lead to corrosion issues.

Therefore, continuous operation of belt dryers is a MUST. Nowadays, electronic, mobile data and alarm transfer are state of the art, meaning that continuous operation can be organized, even if personnel are not present on site 24/7. In the case of drying system failure, the controlling IT shuts down the dryer in a regular manner to achieve a safe condition, and the operator(s) is (are) informed on their mobile phone(s). The energy supply must be constant throughout the period of operation with regard to temperature level and energy amount to ensure even drying and a final DR with minor deviations.

13.3 Corrosion Issues

Incorrect material selection, a dryer design promoting dust deposits on horizontal or only slightly angled surface, incorrect commissioning, or operation with frequent shut-downs may result in corrosion issues in the short and medium term.

One must bear in mind that sludge condensate has a pH-value of up to 10, thus precluding the use of pure alumina, zinc-plated steel, or painted mild steel in the dryer design. Grade 304 (1.4307) stainless steel at normal sites and grade 316L (1.4404) at coastal sites are suitable for normal belt drying systems. High chloride (Cl^-) contents in cooling water may even lead to the use of titanium for heat exchangers.

Surface coatings on materials also undergo heavy wear in the case of high dust content in drying air. As a consequence, corrosion occurs when coatings have disappeared. Therefore, only wear-resistant coatings with ceramic particles are able to prevent abrasion and later corrosion.

In any event, condensate forming inside the dryer must be avoided by increasing the airflow or air temperature or by decreasing water evaporation. Frequent start-up and shutdown processes also need to be prevented, since one can never prevent a level of condensation inside the dryer during a certain warm-up period.

Dust deposits, condensate, and microorganisms have a combined effect. Iron sulfide is commonly present in digested sludge originating from a reaction of iron ions with hydrogen sulfide. Dewatered sludge still contains a residue of hydrogen sulfide, which is released during drying into the gas phase. Deposits of sludge dust together with condensate form an ideal breeding ground for microorganisms inside the dryer. The first processing step is controlled by microorganisms:

$$FeS + H_2S \Leftrightarrow FeS_2 + H_2 \tag{13.1}$$

Iron bisulfide (pyrite) is oxidized by oxygen in an aquatic environment:

$$FeS_2 + 3.5\,O_2 + H_2O \Leftrightarrow FeSO_4 + H_2SO_4 \tag{13.2}$$

Iron sulfate and sulfuric acid are formed, and a temperature rise follows due to the exothermal reaction. Sulfuric acid oxidizes iron in steel:

$$2\,Fe + 4\,H_3O^+ + 2\,SO_4^{2-} + O_2 \Leftrightarrow 2\,FeSO_4 + 6\,H_2O \tag{13.3}$$

Iron(II) sulfate is oxidized and iron(III) sulfate is formed, which itself hydrolyzes and results in the formation of rust and, again, sulfuric acid:

$$4\,FeSO_4 + O_2 + 14\,H_2O \Leftrightarrow 2\,Fe_2O_3 \times H_2O + 8\,H_3O^+ + 4\,SO_4^{2-} \tag{13.4}$$

Thus, in a form of reaction cycle, sulfuric acid transforms considerable quantities of iron into rust, while water is consumed. Even normal stainless steel can be affected. Therefore, condensation inside the dryer must be avoided and dryer design should not allow larger deposits of dust inside the dryer, or these deposits need to be regularly removed.

13.4 Deposits of Elementary Sulfur in Dryer and Condensing Units

It is essential to mention that in the case of high airflow in contact with sewage sludge, hydrogen sulfide evaporating from the sludge immediately reacts with oxygen in drying air, thus forming elementary sulfur and water. Sulfur deposits occur at surfaces with tem-

Fig. 13.1 Heavy sulfur deposits in a direct water-sprayed cross-flow condensing unit. (HUBER SE 2016)

peratures below the dew point of air, i.e., in condensation units and non-insulated parts of the dryer or non-insulated compensators in air tubes. The sulfur deposits often have a crystalline shape. Higher production of crystalline sulfur is seen in cases where the incoming wastewater has a low iron content and no iron compounds are added to digestion to transform H_2S to FeS. A typical Fe content for municipal wastewater is around 5 ± 1 mg/L, recalculated for wastewater generation of 120 instead of the indicated 200 L/(P × d). In the US and UK, values of around 1 mg/L are common. Municipal sewage sludge in Germany has an iron content of around 16,000 mg/kg DM (Koppe and Stotzek 1999). Fig. 13.1 shows heavy sulfur deposits on the bottom of a direct spraying condenser of a belt dryer. This problem was solved by adding caustic agents to circulating water and binding H_2S in the form of salts, which were flushed off with used scrubber water.

During digestion, sulfate is reduced to hydrogen sulfide in the presence of hydrogen or ascetic acid:

$$SO_4^{2-} + 4\,H_2 \Leftrightarrow H_2S + 2\,H_2O + 2\,OH^- \tag{13.5}$$

$$SO_4^{2-} + CH_3COOH \Leftrightarrow H_2S + 2\,HCO_4^- \tag{13.6}$$

Thus, one mole sulfate corresponds to one mole hydrogen sulfide or 1 g SO_4^{2-} to 0.355 g H_2S. The portion of undissociated H_2S is shown in Fig. 13.2 (Kroiss 1985). Usual pH-values of digested thin sludge range from 7.2 to 7.8, yielding portions of undissociated hydrogen sulfide of around 10–40%, forming the maximum quantity of gaseous hydrogen sulfide transferred to drying air during the drying of dewatered sludge.

The pH-dependent portion p of undissolved H_2S in digested thin sludge can be calculated at a temperature of 25 °C with:

$$p(\%) = -4.817 \times \mathrm{pH}^4 + 150.26 \times \mathrm{pH}^3 - 1{,}734.8 \times \mathrm{pH}^2 + 8{,}751.4 \times \mathrm{pH} - 16{,}178 \tag{13.7}$$

with $R^2 = 0.9995$

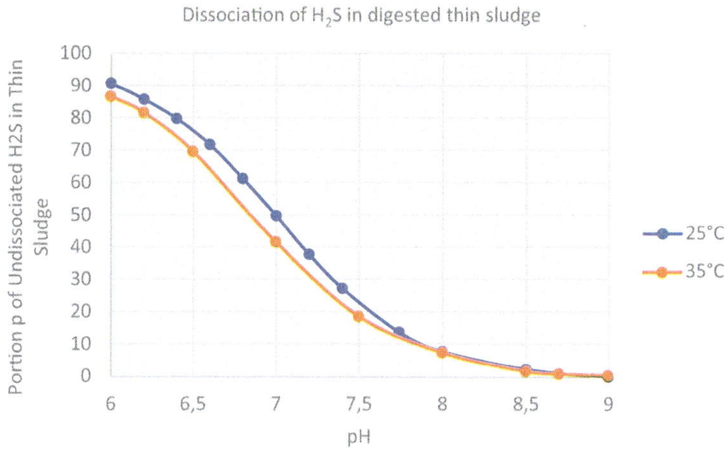

Fig. 13.2 Portion (p) of undissociated H_2S in thin sludge (difference to 100% = portion of HS^-). (According to data from Kroiss 1985 and Yongsiri et al. 2004)

For 35 °C, a common temperature for normal digestion, the equation is:

$$p(\%) = -4.4004 \times \mathrm{pH}^4 + 135.57 \times \mathrm{pH}^3 - 1{,}544.6 \times \mathrm{pH}^2 + 7{,}680.8 \times \mathrm{pH} - 13{,}971 \quad (13.8)$$

with $R^2 = 0.9996$

The balance of undissociated H_2S in thin sludge in digestion with the H_2S in biogas leaving the digester can be described by the following equations and is shown in Fig. 13.3, also viewing temperature dependence (Kroiss 1985):

$$35\,°\mathrm{C:} \quad c_{\mathrm{H_2S\,l\,undiss}}\,(\mathrm{mg/L}) = 26.67 \times c_{\mathrm{H_2S\,g}}\,(\mathrm{vol\%}) \quad (13.9)$$

$$38\,°\mathrm{C:} \quad c_{\mathrm{H_2S\,l\,undiss}}\,(\mathrm{mg/L}) = 24.93 \times c_{\mathrm{H_2S\,g}}\,(\mathrm{vol\%}) \quad (13.10)$$

$$40\,°\mathrm{C:} \quad c_{\mathrm{H_2S\,l\,undiss}}\,(\mathrm{mg/L}) = 22.53 \times c_{\mathrm{H_2S\,g}}\,(\mathrm{vol\%}) \quad (13.11)$$

Thus, by measuring H_2S content in biogas (1 vol% = 10,000 ppm v/v) the maximum possible H_2S content in thin sludge leaving the digester can be rated.

The following chemical agents are suitable to reduce sulfur deposits in belt drying systems—especially at cooler spots where the condensation of water vapor occurs—by dosing into the digester or in front of the digester-feeding pump:

- $FeCl_2$, which is advantageous for direct dosing into the digester due to the fact that less anions and less acid are transferred to the digestion chamber. $FeCl_2$ also causes fewer side reactions in comparison to other chemical agents. No change in pH and no formation of elementary sulfur. Liquid $FeCl_2$ solutions contain 9 or 13 mass% of Fe.

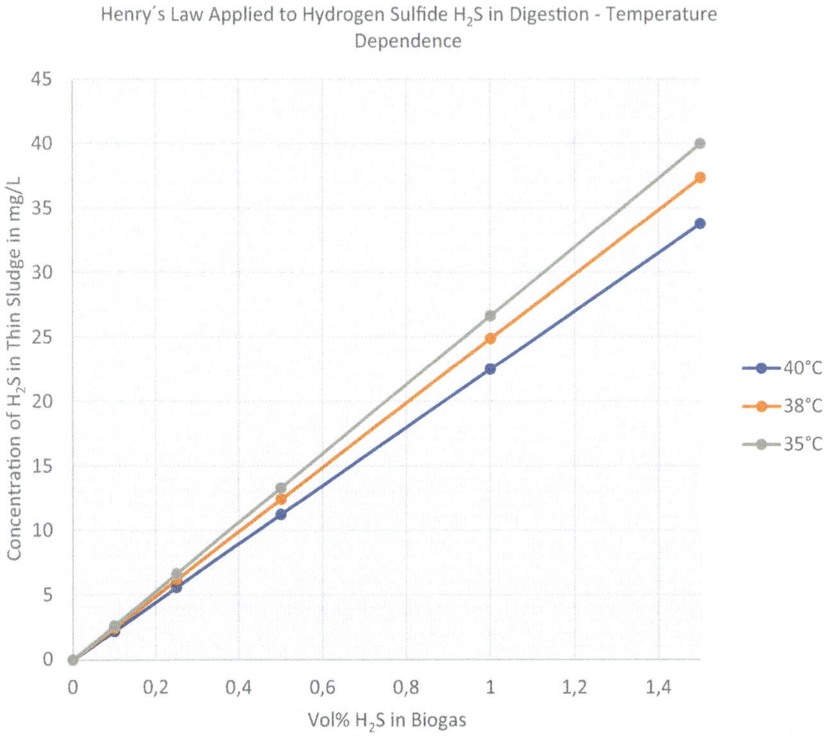

Fig. 13.3 Temperature dependence of balance of H_2S in biogas and thin sludge (undissociated). (Kroiss 1985)

The typical reactions can be written as:

$$Fe_{2+} + 2\,OH^- \Leftrightarrow Fe(OH)_2 \tag{13.12}$$

$$Fe(OH)_2 + H_2S \Leftrightarrow FeS\,(\text{precipitates}) + 2\,H_2O. \tag{13.13}$$

- $FeCl_3$ is more readily available worldwide, but causes more side reactions and needs to be reduced to bivalent iron. Typical reactions are described by:

$$Fe^{3+} + 3\,OH^- \Leftrightarrow Fe(OH)_3 \tag{13.14}$$

$$Fe^{2+} + 2OH^- \Leftrightarrow Fe(OH)_2 \tag{13.15}$$

$$2\,Fe(OH)_3 + H_2S \Leftrightarrow 2\,Fe(OH)_2 + H_2O + S$$
$$(\text{intermediate reaction in digestion, elementary sulfur is formed}) \tag{13.16}$$

$$Fe(OH)_2 + H_2S \Leftrightarrow FeS + 2\,H_2O\,(\text{precipitation reaction in digestion}) \tag{13.17}$$

$$2\,Fe(OH)_3 + 3\,H_2S \Leftrightarrow Fe_2S_3\,(\text{precipitates}) + 6\,H_2O \tag{13.18}$$

$$2\,Fe_2S_3 + 3\,O_2 \Leftrightarrow 2\,Fe_2O_3 + 3\,S$$
$$\text{or}\quad Fe_2S_3 \Leftrightarrow FeS + S \text{ (elementary sulfur precipitates)} \tag{13.19}$$

In addition, HCl is formed in a side reaction and causes a decrease in pH-value depending on the buffering capacity of thin sludge in the digester:

$$2\,FeCl_3 + 3\,H_2S \Leftrightarrow 2\,FeS + 6\,HCl \tag{13.20}$$

- $FeSO_4$ or $FeClSO_4$, which are rarely used, as there is a risk of reforming H_2S out of Na_2SO_4 after a retention time of 30 days in the digester and resolving of sulfur
- FeOH, which is also a possible chemical agent, but far less reactive in comparison to iron chloride compounds

When liquid iron-chloride compounds are used, it is important to bear in mind that they have a high corrosive potential. Therefore, the feeding tube needs to be made of plastic measuring 1.5 m in front of and 5 m behind the feeding point. Feeding time is synchronized with feeding of the digester, and dosing must be finished 5–10 s before the dosing of wastewater to the digester is finished.

Assuming that the H_2S concentration in biogas reaches 2500 ppm (v/v), around 10–15 g of Fe need to be applied per m^3 of biogas to sufficiently reduce H_2S in biogas and sludge. Overdosing of iron with a β-factor of 2.2 is common.

Another option is to add NaOH and H_2O_2 to the circulating water of the direct spraying condenser, thus binding sulfur as a salt, which can be flushed out with the condensing water. A wastewater treatment plant operating 50/100 per year with dosing of chemical agents reported no adverse effect on digestion or generally on wastewater treatment with regard to Na_2SO_4.

13.5 Acceptance of a Planned Sludge Drying Plant

Acceptance by all participants and the local population is a key issue for the successful implementation of a sludge drying project. Therefore, it is essential that the local population, the local authorities including local fire services, and the foreseen personnel are informed in a timely manner of all important issues related to a project of this kind, such as:

- Processing of local sludge to reduce sludge transportation to other regions or countries
- Recovery of phosphorus as an important future task
- Truck traffic associated with the transportation of dewatered sludge
- Cleaning of condensate if no sufficient wastewater treatment is available on site
- Exhaust air cleaning according to the state of the art, taking into consideration the location of the site with regard to the prevailing wind direction and distance to the nearest residential area

- Operator obligations
- Safety issues with regard to possible fire hazards

References

HUBER SE (2016) Photo of a Direct Water-Sprayed Cross-Flow Condensing Unit

Koppe P, Stotzek A (1999) Municipal Wastewater (in German), 4th edn. Vulkan, Essen

Kroiss H (1985) Anaerobic Wastewater Cleaning (In German). Wiener Mitteilungen 62

Yongsiri C et al (2004) Effect of Temperature of Air Water Transfer of Hydrogen Sulfide. J Env Eng 130(1):104–109

14 Legal Framework and Economics

14.1 Legal Framework

In individual countries or communities, the legal framework for the erection and operation of a belt drying plant for sludge must be observed. Official authorization deals with:

- Emissions guidelines
- Storage of dewatered and dried sludge
- Drying of sludge
- Technical instructions on exhaust air purification and noise protection
- Wastewater Feeding Acts for sites without special wastewater treatment facilities
- Water Management Acts
- Ordinance on industrial safety and health
- Explosion protection guidelines
- Fire safety regulations
- Local regulations on development
- Civic involvement, depending on the size of the drying plant

The punctual processing of all these points will prevent considerable loss of time in the completion of a sludge drying project.

A. Heindl, *Belt Drying of Sludge*, https://doi.org/10.1007/978-3-662-66512-1_14

14.2 Economics

The costs of sewage sludge drying—calculated from the sum of capital expenditure (CAPEX) and operational expenditure (OPEX)—are determined by the following factors:

- CAPEX
  - Cost of depreciation (e.g., 7–10% p. a. for machines and 2% p. a. for buildings)
  - Cost of interest
  - Cost of insurance (usually less than 0.5% of the investment)
- OPEX
  - Cost of electrical and thermal (waste) energy (up to 40% of total costs)
  - Costs for skilled personnel for 7/24 operation (up to 5% of total costs)
  - Costs for chemical agents (e.g., 10–30% of total costs for exhaust air treatment including final filtration stage)
  - Cost of material for activated carbon filter or organic material for biofilters
  - Costs for cleaning of condensate in the case of drying units with a condenser (will be far less on a wastewater treatment plant in comparison to other sites)
  - Cost of disposal of dried sludge (if any, fully dried sludge is sometimes accepted without costs)

In some regions, dried sludge may be calculated as revenue since it can be used as a fuel and base material for the cement industry or as control fuel for mono-incineration in fluidized bed combustors. Dried sludge is used as a fertilizer as long as its hygienic status is ensured in some countries such as the US, Middle East countries, and countries in South America.

For German conditions, the total costs ranged from approximately € 50 to € 80/mt of dewatered sludge based on 2018 prices, mainly depending on sludge throughput. Very small units for processing of dewatered sludge of less than 3000–5000 mt/year may result in total costs exceeding € 100/mt. However, the CAPEX and OPEX need to be calculated for each individual project to obtain reliable data for an economic evaluation.

For the calculation of the economics of sludge drying, the charges for disposal of dewatered sludge must be taken into account. If the dewatered sludge can be used as fertilizer, lower disposal costs are incurred. In countries where fertilizer use is restricted, far higher costs need to be taken into account.

Analysis of Wastewater, Sludge, and Dewatered Sludge 15

The analysis of wastewater, thin sludge, and dewatered sludge are important to explain the effects of sludge quality on treatment steps such as dewatering, drying, and incineration of dried sludge. There are national and European standards and collections of analysis methods.

15.1 Measuring the pH-Value of Dewatered Sludge

This measurement is performed according to EN 15933 (2012) Sludge, treated biowaste and soil—Determination of pH. Dewatered sludge, which has a moisture content of 70–80% relative to total mass, needs to be thinned with five parts by mass of distilled water before pH measurement with common pH electrodes can be carried out. Water quality must be 2 according to EN ISO 3696, i.e., electrical conductivity $< 0.2\,mS/m$ at 25 °C and $pH \leq 5.6$. Of course, the thinning of sludge has an effect on pH-value. The dissociation of water around the pH-value of 7 also has a pronounced influence. To obtain the pH-value of the dewatered sludge, the measured value needs to be corrected. Fig. 15.1 shows that the actual pH value below $pH = 7$ is lower and above $pH = 7$ higher than measured in the thinned sample. No correct value can be assigned around the pH-value of 7.

A. Heindl, *Belt Drying of Sludge*, https://doi.org/10.1007/978-3-662-66512-1_15

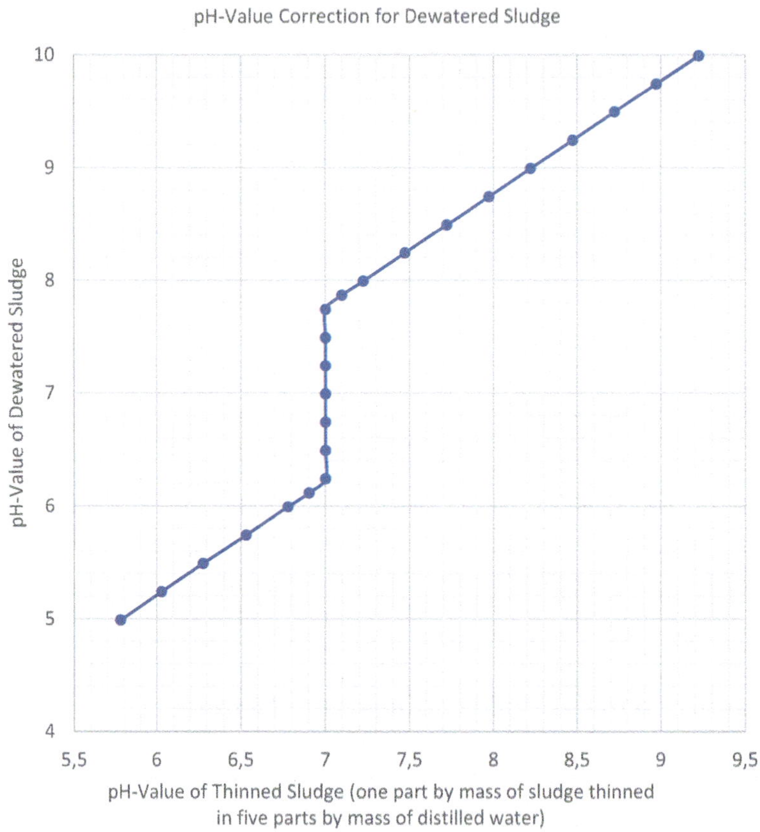

Fig. 15.1 Corrected pH-value of dewatered sludge

15.2 Measurement of Respiration Activity

The respiration activity is analyzed to evaluate the degree of stabilization of aerobically stabilized sludge. The following procedural instructions can be given:

- Take 0.5 L sludge with a known dry substance (DS) content from the outlet of the aeration system, heat it to around 20 °C (water bath) and shake the sludge in a 1-L bottle for at least 20–30 s for O_2 enrichment to over 6 mg/l.
- Use a Winkler bottle or Karlsruher bottle filled with 2 ml 0.05% N-allylthiourea (ATH) solution (200–300 ml; see equipment description below), fill it with aerated sludge up to the rim and place in magnetic bars. ATH 0.05% solution: dissolve 0.25 g ATH in 500 mL distilled water.
- Fill a Winkler bottle or Karlsruher bottle (200–300 ml) with aerated sludge up to the rim and place in magnetic bars.

- Put the bottle in a water bath and place this on a magnetic stirrer (e.g. plastic cup with about 20° warm water). The temperature must not deviate more than 2 °C from the nominal temperature of 20 °C. If a larger thermo-cabinet is available, the measurement can be conducted there. Optionally, a magnetic stirrer from a biochemical oxygen demand (BOD) measuring device may be suitable to create turbulence.
- Insert the O_2 probe of the calibrated O_2 device (handheld or laboratory device) up to the middle of the bottle (do not entrap air!) and increase the stirrer speed to such a level that the flow turbulence required for the probe is achieved.
- After a lead time of 5–10 min, the O_2 concentration is measured every 1–3 min depending on the velocity of O_2 depletion. The measurements are recorded. The recorded measurements can be ended after about 15 min, but at the latest when the O_2 concentration has reached 0.5 mg/L. Lower values are disregarded when calculating activity.

Sludge respiration activity is calculated as g O_2 consumption per 1 kg total solids (TS) and per day on the basis of the following formula:

$$O_2 - \text{respiration activity} = \frac{O_2 - \text{consumption}}{\text{TS}} \tag{15.1}$$

Example: TS of the sludge, 4.8 g/L; O_2 at the beginning of recording, 4.1 mg/L; O_2 after 15 min (end of measurements), 1.5 mg/L; O_2 depletion per minute of 0.173 mg/(L × min):

$$O_2 - \text{respiration activity} = 0.173\,\frac{\text{mg}}{\text{L} \times \text{min}} \times 1440 \times \frac{1}{4.8\,\frac{\text{g}}{\text{L}}} = 52\,\frac{\text{g}}{\text{kgTS} \times \text{d}} \tag{15.2}$$

Evaluation:

- If the calculated respiration activity is below 100 g/(kg TS × d), the tested sludge is sufficiently (limitedly) aerobically stabilized sludge.
- If the calculated activity is higher, the sludge is only partially aerobically stabilized. Only at values as low as below 50 g/(kg TS × d) is the sludge well aerobically stabilized.

Equipment:

- Oxygen bottles (so-called Winkler bottles), 200–300 mL content (exact bottle volume specification not required). The neck of the bottle needs to be adjusted to suit the O_2 laboratory probe used. The Winkler bottle, better known as an oxygen bottle according to L. S. Winkler, consists of glass and serves for the determination of the oxygen dissolved in water. The measured volume is indicated accurately to ±0.01 mL. There is a labeling field on the bottle and the bottle has a massive, conical standard ground glass stopper, which can be secured with a spring clip. Each bottle is adjusted with its own stopper. The stopper and bottle are therefore not interchangeable and must always be marked with numbers to avoid any confusion. The nominal volumes are 100–150 mL with 14/23 standard ground joint and 250–300 mL with 19/26 standard ground joint (Wikipedia 2016).

- Oxygen measuring device with laboratory O_2 probe, preferably with digital display
- Magnetic stirrer with infinitely variable rotary speed, preferably 12 V operation (adapter required)
- Magnetic stir bars, at least 30 mm long, e.g., stir bars of BOD measuring devices
- mL pipette
- Annealing furnace (to determine the loss on ignition *GV* or concentration of the organic dry substance TS or the concentration of the organic dry mass volatile total solids [VTS])

15.3 Determination of Alkalinity

Alkalinity is used to characterize the degree of stabilization of biological sludge (waste activated sludge).

Titration is carried out using a standard acid solution up to the fixed pH endpoints 8.3 and 4.5. The endpoints can be determined visually or photometrically. They are in accordance with selected equivalence points for the determination of hydrogen carbonate, carbonate, and hydroxide. At the pH value of 8.3, the equivalence concentration of carbonate and carbon dioxide is reached, corresponding to the titration of the total present hydroxide and half the carbonate. At the pH-value of 4.5, the equivalence concentration of hydrogen ions and hydrogen carbonate ions is reached, once all bases have been titrated. This is the total alkalinity of the sample. The details of the procedure are described in the European standards EN ISO 9963-1, which are also available in English.

15.4 Different Regulations for Sludge Analysis

There are different European and German regulations for sludge analysis:

- DIN EN 12880:2001-02 Characterization of sludge—Determination of dry residue and water content (= 1-DR)
- DIN EN 14671:2006-09 Characterization of sludge—Pre-treatment for the determination of extractable ammonia using 2 mol/l potassium chloride
- EN 15170:2009-05 Characterization of sludge—Determination of calorific value
- EN 15170 (2008) Characterization of sludge determination of calorific value
- EN 15934:2012-11 Sludge, treated biowaste, soil and waste—Calculation of dry matter fraction after determination of dry residue or water content
- EN 15935 (2012) Sludge, treated biowaste, soil and waste—determination of loss on ignition

- EN ISO 17892-4:2017-04 Geotechnical investigation and testing—Laboratory testing of soil. Part 4: Determination of particle size distribution (ISO 17892-4:2016, sieves used with mesh sizes in mm: 10, 4, 2, 1, 0.5, 0.25, 0.1, 0.05, <0.05. Sieving time 30 min, sieving of dried sludge)
- DIN 66165-2:2016-08 Particle size analysis—Sieving analysis. Part 2: Procedure 2016-08, sieving of dewatered sludge

In the US, detailed information on wastewater analysis can be found at:

- https://www.standardmethods.org/doi/book/10.2105/SMWW.2882: Baird R B Bridgewater L (2017) Standard Methods for the Examination of Water and Wastewater, 23rd Edition by R.B. Baird, American Water Works Association (AWWA, WEF and APHA)
- EPA Clean Water Act Analytical methods https://www.epa.gov/cwa-methods
- Code of Federal Regulations 40 CFR Part 136 https://www.ecfr.gov/current/title-40/chapter-I/subchapter-D/part-136?toc=1

Definitions and Basic Explanations 16

BOD_5 Biological oxygen demand: Quantity of oxygen consumed in 5 days by wastewater or thin sludge after incubation. Nitrification reactions need to be blocked to obtain the exact value of biodegradable organic carbon pollution.

COD Chemical oxygen demand: Overall consumption of oxygen in potassium dichromate, covering the greater portion of organic compounds and mineral salts being oxidized.

Biosolids Stabilized sludge meeting the criteria of the US EPA 40 CFR 503 regulations that can be used, for example, as fertilizer. The stabilization process can be carried out by: aerobic or anaerobic digestion; alkaline treatment using, for example, lime and increasing pH-value; composting at increasing temperature; or thermal drying decreasing water content and increasing temperature.

Sludge Any residue produced during primary, secondary, or advanced wastewater treatment that has not been treated to reduce pathogens or vector attraction. The term 'sludge' is further defined by a process descriptor such as primary sludge, secondary sludge = waste activated sludge or biological sludge, mixed sludge, aerated sludge, or digested sludge. Hydrolyzed sludge is characterized by thermal or thermochemical treatment of sludge to rupture cell structure and increase dewaterability. The term sludge is often also used for biosolids. In fact, after wastewater treatment, sludge has a certain degree of stabilization.

Stabilization Treatment of solids generated in wastewater cleaning to convert them to a stable product, often simultaneously reducing pathogens and odor development (WEF 2010).

Municipal sludge It is generally known that municipal wastewater treatment plants also handle industrial wastewater. As long as the mass portion of industrial wastewater is less than 30%—typically 5–25% depending on plant size and industrial settlements—the generated sludge can be classified as municipal sludge. However, values exceeding 30% result in a municipal-industrial sludge with varying qualities deter-

A. Heindl, *Belt Drying of Sludge*, https://doi.org/10.1007/978-3-662-66512-1_16

mined by the flux of industrial wastewater. Sludge of this kind is also encountered when a high proportion of external co-substrates is used to boost biogas generation.

Sticky phase of sludge During drying, water is removed and dry residue (DR) increased. In thermal drying, the sticky phase (SP) can be encountered in a DR range of 40–45% to 60–65%. The sludge sticks to surfaces and becomes sticky and pasty, larger agglomerates are formed, and its suitability for mechanical conveyance deteriorates dramatically, whereas the power demand for mechanical conveyance increases enormously. Non-stabilized or poorly stabilized sludge with high volatile dry residue and protein content tends to reach the sticky phase at lower DR of around 40%. In low temperature solar drying of sludge with prolonged drying time and mechanical mixing, the SP occurs at DR contents of 15–25%, i.e., at far lower DR contents (DWA 2021). The stickiness is strongly related to extracellular polymeric substances (e.g., proteins) and their concentration in the sludge matrix. With decreasing water content and increasing total solid content, the long biopolymers concentrate, glue together, and incorporate the microorganisms, thereby creating a sticky, pasty sludge mass (Peeters et al. 2014). Mechanical stress on sludge further increases stickiness.

Dust The term for particles in dried sewage sludge with a diameter of less than 500 μm and posing a risk of self-ignition if agglomerations in the dryer are not readily removed or a risk of dust explosion. Excessive drying to more than 92% DR or drying with partial back-mixing of dried material results in higher dust contents in drying air.

Log K_{OW} Octanole–water partition coefficient of gases; lower values result in a higher water solubility of a substance, while higher values result in a higher solubility in fatty substances.

People's values People's values (P) are expressed as the sum of the population whose wastewater is treated in a wastewater treatment plant plus an additional people's equivalent value (PE; calculated by BOD_5 load from industrial wastewater divided by 60). For example, the production of vegetable or fruit conserves creates a PE value of 1 per 2 kg of vegetable conserves. Thus, in a town of 100,000 inhabitants, the P value of the wastewater treatment plant receiving the wastewater from a vegetable processing plant of 100 t per year is calculated to be 150,000 PE.

VOC Volatile organic compound, describing gas components based on carbon (C) that are transferred to exhaust air during drying and expressed in mg C/m^3.

References

DWA (2021) Sewage sludge Drying (In German). DWA-Regelwerk Merkblatt DWA-M 379. DWA (Deutsche Vereinigung für Wasserwirtschaft, Abwasser und Abfall e.V.), Hennef

Peeters B et al (2014) Challenges of Drying Sticky Wastewater Sludge. https://www.chemengonline.com/challenges-of-drying-sticky-wastewater-sludge/?printmode=1. Accessed 22 Aug 2021

WEF (2010) Design of municipal waste water treatment plants. Volume 3 Solids Processing and Management, 5th edn. Water Environment Federation WEF. McGraw Hill, New York

Index

A. Heindl, *Belt Drying of Sludge*, https://doi.org/10.1007/978-3-662-66512-1

The manufacturer's authorised representative in the EU is Springer Nature Customer Service Centre GmbH, Europaplatz 3, 69115 Heidelberg, Germany. If you have any concerns regarding our products, please contact ProductSafety@springernature.com

Printed and bound by CPI Group (UK) Ltd, Croydon, CR0 4YY
30/07/2026
02182046-0014